*Edited by Angelo Chianese and
Herman J. M. Kramer*

**Industrial Crystallization
Process Monitoring and Control**

Related Titles

Houson, I. (ed.)

Process Understanding

For Scale-Up and Manufacture of Active Ingredients

2011
ISBN: 978-3-527-32584-9

Pollak, P.

Fine Chemicals

The Industry and the Business

Second edition
2011
ISBN: 978-0-470-62767-9

Duffar, T. (ed.)

Crystal Growth Processes Based on Capillarity

Czochralski, Floating Zone, Shaping and Crucible Techniques

2010
ISBN: 978-0-470-71244-3

Tiekink, E. R. T., Vittal, J., Zaworotko, M. (eds.)

Organic Crystal Engineering

Frontiers in Crystal Engineering

2010
ISBN: 978-0-470-31990-1

Capper, P., Rudolph, P. (eds.)

Crystal Growth Technology

Semiconductors and Dielectrics

2010
ISBN: 978-3-527-32593-1

Tung, H.-H., Paul, E. L., Midler, M., McCauley, J. A.

Crystallization of Organic Compounds

An Industrial Perspective

2009
ISBN: 978-0-471-46780-9

Hessel, V., Renken, A., Schouten, J. C., Yoshida, J.-i. (eds.)

Micro Process Engineering

A Comprehensive Handbook

3 Volume Set
2009
ISBN: 978-3-527-31550-5

Scheel, H. J., Capper, P. (eds.)

Crystal Growth Technology

From Fundamentals and Simulation to Large-scale Production

2008
ISBN: 978-3-527-31762-2

Edited by Angelo Chianese and Herman J. M. Kramer

Industrial Crystallization Process Monitoring and Control

WILEY-VCH Verlag GmbH & Co. KGaA

The Editors

Prof. Dr. Ing. Angelo Chianese
Sapienza University of Rome
Department of Chemical Engineering
Materials Environment
via Eudossiana 18
00184 Rome
Italy

Prof. Dr. Herman J. M. Kramer
Delft University of Technology
Process & Energy Laboratory
Leeghwaterstraat 44
2628 CA Delft
The Netherlands

■ All books published by **Wiley-VCH** are carefully produced. Nevertheless, authors, editors, and publisher do not warrant the information contained in these books, including this book, to be free of errors. Readers are advised to keep in mind that statements, data, illustrations, procedural details or other items may inadvertently be inaccurate.

Library of Congress Card No.: applied for

British Library Cataloguing-in-Publication Data
A catalogue record for this book is available from the British Library.

Bibliographic information published by the Deutsche Nationalbibliothek
The Deutsche Nationalbibliothek lists this publication in the Deutsche Nationalbibliografie; detailed bibliographic data are available on the Internet at <http://dnb.d-nb.de>.

© 2012 Wiley-VCH Verlag & Co. KGaA, Boschstr. 12, 69469 Weinheim, Germany

All rights reserved (including those of translation into other languages). No part of this book may be reproduced in any form – by photoprinting, microfilm, or any other means – nor transmitted or translated into a machine language without written permission from the publishers. Registered names, trademarks, etc. used in this book, even when not specifically marked as such, are not to be considered unprotected by law.

Cover Design   Formgeber, Eppelheim
Typesetting   Laserwords Private Limited, Chennai, India
Printing and Binding   Markono Print Media Pte Ltd, Singapore

Printed in Singapore
Printed on acid-free paper

Print ISBN: 978-3-527-33173-4
ePDF ISBN: 978-3-527-64518-3
oBook ISBN: 978-3-527-64520-6
ePub ISBN: 978-3-527-64517-6
Mobi ISBN: 978-3-527-64519-0

Contents

Preface *XI*
Scope of the Book *XIII*
List of Contributors *XXI*

1 **Characterization of Crystal Size Distribution** *1*
Angelo Chianese
1.1 Introduction *1*
1.2 Particle Size Distribution *1*
1.3 Particle Size Distribution Moments *4*
1.4 Particle Size Distribution Characterization on the Basis of Mass Distribution *5*
References *6*

2 **Forward Light Scattering** *7*
Herman J.M. Kramer
2.1 Introduction *7*
2.2 Principles of Laser Diffraction *8*
2.3 Scatter Theory *10*
2.3.1 Generalized Lorenz–Mie Theory *12*
2.3.2 Anomalous Diffraction *12*
2.3.3 Fraunhofer Diffraction *13*
2.4 Deconvolution *13*
2.4.1 Direct Inversion Using the Nonnegativity Constraint *14*
2.4.2 Philips Twomey Inversion Method *14*
2.4.3 Iterative Methods *15*
2.5 The Effects of Shape *15*
2.6 Multiple Scattering *16*
2.7 Application of Laser Diffraction for Monitoring and Control of Industrial Crystallization Processes *17*
2.8 Conclusions *19*
References *20*
Further Reading *20*

3		**Focused Beam Reflectance Measurement** *21*
		Jochen Schöll, Michel Kempkes, and Marco Mazzotti
3.1		Measurement Principle *21*
3.2		Application Examples *21*
3.2.1		Solubility and Metastable Zone Width (MSZW) *21*
3.2.2		Seed Effectiveness *22*
3.2.3		Polymorph Transformations *22*
3.2.4		Effect of Different Impurity Levels *23*
3.2.5		Nucleation Kinetics *24*
3.2.6		Improved Downstream Processing *24*
3.2.7		Process Control *25*
3.3		Advantages and Limitations *26*
		References *27*
4		**Turbidimetry for the Estimation of Crystal Average Size** *29*
		Angelo Chianese, Mariapaola Parisi, and Eugenio Fazio
4.1		Introduction *29*
4.2		Determination of Average Particle Size from Specific Turbidity *29*
4.3		Procedure to Evaluate Average Crystal Size by Turbidimetry for a High Solid Slurry Concentration *31*
4.4		Conclusion *34*
		References *34*
		Further Reading *34*
5		**Imaging** *35*
		Herman J.M. Kramer and Somnath S. Kadam
5.1		Introduction *35*
5.2		Literature Overview *36*
5.3		The Sensor Design *39*
5.3.1		Optics and Illumination *40*
5.3.2		The Camera System and the Resolution *42*
5.3.3		Image Analysis *43*
5.3.4		Statistics *46*
5.4		Application of In Situ Imaging for Monitoring Crystallization Processes *46*
5.4.1		Example 1 *46*
5.4.2		Example 2 *47*
5.5		Conclusions *48*
		References *49*
		Further Reading *50*
6		**Turbidimetry and Nephelometry** *51*
		Angelo Chianese, Marco Bravi, and Eugenio Fazio
6.1		Introduction *51*
6.2		Measurement of Nucleation and Solubility Points *51*

6.3	The Developed Turbidimetric and Nephelometric Instruments	52
6.4	The Examined Systems	53
6.5	Obtained Results	54
	References	57

7 Speed of Sound   59
Joachim Ulrich and Matthew J. Jones

7.1	Introduction	59
7.2	In-Process Ultrasound Measurement	59
7.3	Determining Solubility and Metastable Zone Width	60
7.4	Measuring Crystal Growth Rates	65
7.5	Detecting Phase Transitions with Ultrasound	66
	References	68

8 In-Line Process Refractometer for Concentration Measurement in Sugar Crystallizers   71
Klas Myréen

8.1	Introduction	71
8.2	Measurement Principle	72
8.3	In-Line Instrument Features and Benefits	74
8.3.1	Accuracy	74
8.3.2	Concentration Determination	74
8.3.3	Process Temperature Compensation Factor	75
8.3.4	Process Sensor	75
8.4	Features and Benefits	76
8.5	Example of Application in the Crystallization	76
8.5.1	Seeding Point and Supersaturation Control in Sugar Vacuum Pan	77
8.6	Conclusion	79
	References	79

9 ATR-FTIR Spectroscopy   81
Christian Lindenberg, Jeroen Cornel, Jochen Schöll, and Marco Mazzotti

9.1	Introduction	81
9.2	Calibration	82
9.3	Speciation Monitoring	84
9.4	Co-Crystal Formation	84
9.5	Solubility Measurement	86
9.6	Crystal Growth Rates	86
9.7	Polymorph Transformation	87
9.8	Crystallization Monitoring and Control	89
9.9	Impurity Monitoring	89
9.10	Conclusions	90
	References	90

10	**Raman Spectroscopy** 93	

Jeroen Cornel, Christian Lindenberg, Jochen Schöll, and Marco Mazzotti

- 10.1 Introduction 93
- 10.2 Factors Influencing the Raman Spectrum 94
- 10.3 Calibration 95
- 10.3.1 Univariate Approaches 95
- 10.3.2 Multivariate Approaches 98
- 10.4 Applications 99
- 10.4.1 Solid-Phase Composition Monitoring 99
- 10.4.2 Liquid Phase Composition Monitoring 99
- 10.4.3 Amorphous Content Quantification 100
- 10.5 Conclusions 101
- References 102

11 Basic Recipe Control 105

Alex N. Kalbasenka, Adrie E.M. Huesman, and Herman J.M. Kramer

- 11.1 Introduction 105
- 11.2 Incentives for Basic Recipe Control 105
- 11.3 Main Mechanisms, Sensors, and Actuators 106
- 11.3.1 Crystallization Mechanisms 106
- 11.3.2 Sensors 106
- 11.3.3 Measurement of the Solute Concentration 107
- 11.3.4 Measurement of the Crystal Number, Size, Distribution, and Morphology 107
- 11.3.5 Actuators 108
- 11.4 Basic Recipe Control Strategy 109
- 11.4.1 How to Obtain a Recipe? 110
- 11.4.2 Scaling Up the Recipe 111
- 11.5 Seeding as a Process Actuator 111
- 11.5.1 Initial Supersaturation 112
- 11.5.2 Seed Mass 112
- 11.5.3 Seed Size and Size Distribution 113
- 11.5.4 Seed Quality and Preparation Procedure 113
- 11.5.5 Methods of Addition of Seeds 114
- 11.6 Rate of Supersaturation Generation 114
- 11.7 Mixing and Suspension of Solids 116
- 11.8 Fines Removal and Dissolution 118
- 11.9 Implementation of Basic Recipe Control 119
- 11.10 Conclusions 122
- References 122

12 Seeding Technique in Batch Crystallization 127

Joachim Ulrich and Matthew J. Jones

- 12.1 Introduction 127
- 12.2 Seeding Operation: Main Principles and Phenomena 127

12.3	Use of Seeding for Batch Crystallization: Main Process Parameters *128*	
12.4	Control of Batch Crystallization by Seeding: Empirical Rules for Design *131*	
	References *137*	
13	**Advanced Recipe Control** *139*	
	Alex N. Kalbasenka, Adrie E.M. Huesman, and Herman J.M. Kramer	
13.1	Introduction *139*	
13.2	Incentives and Strategy of the Advanced Recipe Control *139*	
13.3	Modeling for Optimization, Prediction, and Control *141*	
13.4	Model Validation *143*	
13.5	Rate of Supersaturation Generation *144*	
13.6	Mixing Conditions *145*	
13.7	Implementation *150*	
13.8	Example of Modeling, Optimization, and Open-Loop Control of a 75-l Draft-Tube Crystallizer *150*	
13.8.1	Objectives and Advanced Recipe Control *150*	
13.8.2	Process Description and Modeling *151*	
13.8.3	Dynamic Optimization *153*	
13.8.4	Experimental Validation Results *155*	
13.9	Conclusions *157*	
	References *158*	
14	**Advanced Model-Based Recipe Control** *161*	
	Alex N. Kalbasenka, Adrie E.M. Huesman, and Herman J.M. Kramer	
14.1	Introduction *161*	
14.2	Online Dynamic Optimization *163*	
14.3	MPC for Batch Crystallization *168*	
14.4	Conclusions and Perspectives *172*	
	References *173*	
15	**Fines Removal** *175*	
	Angelo Chianese	
15.1	Introduction *175*	
15.2	Fines Removal by Heat Dissolution *175*	
15.3	Modeling of an MSMPR Continuous Crystallizer with Fines Removal *177*	
15.4	Fines Destruction in the Industrial Practice *178*	
15.5	CSD Control by Fines Removal for Pilot Scale Crystallizers *180*	
15.6	The Cycling Phenomenon as Undesired Effect of Fines Destruction in Industrial Crystallizers *182*	
	References *184*	

16	**Model Predictive Control** *185*	
	Alex N. Kalbasenka, Adrie E.M. Huesman, and Herman J.M. Kramer	
16.1	Introduction *185*	
16.1.1	Receding Horizon Principle *185*	
16.1.2	Advantages and Disadvantages of MPC *187*	
16.2	Approach for Designing and Implementing an MPC Control System *189*	
16.3	Process Modeling *190*	
16.4	The Performance Index *192*	
16.5	Constraints *193*	
16.6	The MPC Optimization *193*	
16.7	Tuning *193*	
16.8	State Estimation *194*	
16.9	Implementation *196*	
16.10	MPC of Crystallization Processes *197*	
16.11	Delta-Mode MPC *198*	
16.12	Conclusions and Perspectives *199*	
	References *200*	
17	**Industrial Crystallizers Design and Control** *203*	
	Franco Paroli	
17.1	Introduction *203*	
17.2	Forced Circulation Crystallizer *204*	
17.3	Draft-Tube-Baffle Crystallizer *209*	
17.3.1	"Oslo" Growth-Type Crystallizer *213*	
17.4	Process Variables in Crystallizer Operation *218*	
17.4.1	Continuous Operation *218*	
17.4.2	Batch Operations *219*	
17.5	Sensors *219*	
17.5.1	Level *220*	
17.5.2	Density *220*	
17.5.3	Crystal Size *221*	
17.6	Control Devices *222*	
	References *224*	

Index *225*

Preface

The idea of this book was to disseminate some valuable results achieved by two European projects on monitoring and control of industrial crystallizers:

1) CRYSEN (2000–2003) on the Development of new sensors for industrial crystallization;
2) SINC-PRO (2003–2005) on Self-learning model for INtelligent predictive Control system for crystallization PROcesses. This second European project then became an international project with the partnership extended to Swiss and Japanese teams.

Most of the partners of the two projects were members of the Working Party on Crystallization (WPC) of the European Federation of Chemical Engineers, which accepts the proposal of the book and encouraged the efforts of the two editors in order to provide a new publication to the industrial crystallizer community. Then, the WPC provided a double reviewing of each books chapter by the WPC members, from academia and industry, expert in the specific subject.

Therefore, the two editors are greatly grateful to the two Chairmen of the WPC, who promoted the book writing, Joachim Ulrich and Beatrice Biscans and to the following WPC members, who with their referee's work contributed to improve the quality of the book:

Beatrice Biscans
Colm Campbell
Philippe Carvin
Roger Daudey
Brian Glennon
Laurent Laferrere
Jaime Morales
Zoltan Nagy
Matthias Raouls
Ake Rasmuson

December 2011 *Angelo Chianese and Herman J. M. Kramer*

Scope of the Book

Control of Crystallization Processes in Industrial Practice

Crystallization from a solution is a separation technique, where one of more components of the solution are separated as a solid phase. Application of a crystallization process can be aimed at the separation of a component from a multicomponent mixture, but in most cases it is focused on the production of a solid product from a liquid mixture. In both types of application, the product quality of the solid product has to meet stringent specifications, and especially for solid products manufacturers have to meet ever-increasing demands of the customers on particle properties such as particle size distribution, crystal shape, degree of agglomeration, caking behavior, and purity. Since for an economic beneficial operation a smooth separation of the produced crystals from the mother liquor is essential, additional demands on filterability and washability need to be obeyed. It is obvious that in order to achieve the increasing demands on product quality, crystallization processes have to be carefully controlled.

Crystallization from a liquid solution is the most important production and separation process in the fine chemical and food industry and one of the most important processes in the chemical process industry as a whole. Overall it is estimated that 70% of the products sold by the chemical process industry are solids. Many intermediates (e.g., adipic acid for nylon production), fine chemicals, pharmaceuticals (e.g., aspirin), biochemical, food additives, and bulk products such as fertilizers are solids obtained through crystallization.

Both market needs and governmental policies demand continuous efforts in research and development to improve existing technologies, with respect to economics, operability, and sustainability.

In order to achieve an optimal production capacity and the desired crystal properties, the process conditions during the crystallization operation should be controlled in such a way that the product specifications can be met at conditions of profitability and trouble-free production. The crystallization process variables either should be controlled by the configuration of the process/equipment or must be manipulated during operation, either to achieve a desired profile during batch operation or to compensate for process disturbances during continuous operation. In this respect three different levels of control can be identified. The first level consists of a base layer control to keep constant the basic process variables such as the materials and enthalpy fluxes the gas–liquid level, the temperature and the pressure in the crystallizer at their design values to ensure a trouble-free reproducible production. The next level, the control of supersaturation, is more complicated as it requires a precise monitoring of the solute concentration and precise knowledge of the saturation concentration at the process conditions. The setting of this variable entails detailed knowledge of the optimal operation window for the driving force to achieve an optimal balance between the production capacity and the product quality. Finally, the control of the product quality as a whole, in most cases expressed in terms of its crystal size distribution (CSD), is the third, most difficult and worthwhile level of control and requires both an online monitoring of this product quality and detailed knowledge of the process. The advantage of supersaturation control is that in principle it allows for a so-called model-free control strategy, avoiding time-consuming development and validation of a detailed process model as is required for controlling the product quality directly.

Industrial crystallizers are seldom operated under automatic control schemes because of the lack of reliable sensors and process models. Robust and reliable sensors for *in situ* monitoring of the relevant process variables, that is, the CSD and the solute concentration, were not available for a long time and only recently some *in situ* sensors have started to be applied at industrial level. The development of process models is another obstacle for the application of feedback (FB) schemes for quality control. Due to the absence of general reliable crystallization models and problems encountered in the prediction of the effects of scale up, dedicated models have to be developed and validated for each individual case which has proven to be a very difficult and costly matter. This leads us to adopt open loop quality process control schemes where the main process variable are operated at the conditions that are supposed to lead to the desired product quality, without an automatic closed loop control of product quality. Unfortunately, the absence of an effective control system typically entails a poor quality control, which must be compensated for by additional processing such as follows:

- classification of crystalline product streams and reprocessing of under and oversized (agglomerated) products;
- exhaust air treatment for dust abatement;
- higher energy consumption in the drying process and reprocessing of the off-spec product;
- significant decrease of productivity resulting in reduced total plant throughput and impacting investment costs to reach the required capacity;
- production costs increase (about 5% of total production costs could be avoided adopting an effective control in batch processes); and
- higher environmental impact due to the required increase in use of solvent and chemicals.

Ideally industrial crystallizers are operated in such a way that the product specifications are met under conditions which permit a profitable trouble-free production of the desired crystalline material. In industrial practice however, a large number of operational problems can be encountered which reduce the crystallizer performances, such as deposition of crystalline material on the crystallizer internals, variations in the feed composition, less effective heat transfer operations, inappropriate seeding procedures, and so on. These process disturbances will inevitably lead to production losses and/or deterioration of the product quality.

In conclusion, the problem in the quality control of industrial crystallizers is far from its solution. For this reason, much effort has to be spent on the development of new online sensors and more advanced control approaches.

Different control approaches are applied for crystallization processes belonging to pharmaceutical plants and commodities plants. In the first case, usually processes are performed in batch mode, by applying cooling or antisolvent techniques, concerning high added value products, whose quality is the first objective to be pursued. It is well known that even minor changes in crystallization process conditions and equipment, for example, supersaturation, temperature, impurity, cooling rate, or reactor hydrodynamics, can result in significant variations in the crystal and downstream powder properties, notably, polymorphic form, particle size, shape, purity, and defect structure. The market price of the active products may reach values of several thousands of Euros per kilogram, thus allowing the achievement of sophisticated instrumentations to monitor and control the product quality throughout the crystallization process.

The second area of crystallization applications concerns so-called commodities products, processed in continuous mode in huge amount with relatively poor specifications and having a small added value which does not allow high investment cost per product unity.

The development and application in recent years of expansive and sophisticated sensors are due to the increasing interest from the pharmaceutical companies in the improvement of the crystallization processes operation and their large need for research investigations at lab scale. Among these instruments are those based on attenuated total reflection (ATR)-Fourier transform infrared (FTIR) spectroscopy, *in situ* chord length distribution of crystals from laser backscattering by focus beam reflectance measurement (FBRM) probe and *in situ* online video microscope. Most of the available *in situ* sensors are robust enough to be applied in the production environment. This opened the possibility of FB control-based crystallization design and operation. The new opportunities are well described by the guideline document issued in 2004 by the U.S. Food and Drug Administration (FDA), as part of a broader initiative on current Good Manufacturing Practices (cGMP) (FDA, 2004). This document emphasized the development and use of novel technologies based on process analytical technologies (PAT) as a tool for "twenty-first century manufacturing" moreover, the development of tailored process control strategies was recognized as the most important way to prevent or mitigate the risk of producing poor quality products.

This new scenario provides significant potential for implementation of optimal and adaptive control methodologies with real economic benefits associated with better product quality (Nagy, Fujiwara, and Braatz, 2008; Woo, Tan, and Braatz, 2011).

A quite different situation holds for the crystallization processes included in commodities plants. In fact, in this case the productivity and the CSD are the main issues. In order to achieve these two objectives, the use of traditional online sensors with improvement performances, such as turbidimeters and refractometers, is welcome. Only seldom new sophisticated and expensive instruments such as the FBRM sensor are adopted and most of the quality product assessments in terms of the CSD are carried out off-line by taking samples and making use of sieve analysis, laser diffraction instruments, or an optical microscope. More attention is devoted to the manipulation of specific variables to maintain the crystallizer under control, even if in a nonautomatic way. In this respect the fines removal, the agitation by an impeller, and the amount of added seed crystals are the most used manipulating variables. However, to make use of various features to control the product quality in modern continuous crystallizers, such as draft tube baffled (DTB) crystallizers, simultaneous manipulation of different process inputs is needed. The full exploitation of these crystallizers therefore requires the application of multivariable control techniques, especially when different aspects of the product quality have to be controlled or when the product quality needs to be preserved at different production capacities, as has been shown in several research studies (Trifkovic, Sheikhzadeh, and Rohani, 2009; Sheikhzadeh, Trifkovic, and Rohani, 2008; Seki, Amano, and Emoto, 2010; Valencia Peroni, Parisi, and Chianese, 2010).

Content of the Book

In this book the monitoring and control of industrial crystallizers is discussed. All the necessary ingredients for the development and implementation of a control strategy for batch and continuous operated crystallization processes are reported. The emphasis will be on cooling and evaporative crystallization processes, although the methodology can also be applied for other types of crystallization processes such as pH shift and antisolvent crystallization or precipitation processes. The book is written primarily for process and control engineers interested in improving the performances of their crystallization processes and for chemical and control engineering students interested in application of online sensors and control schemes to crystallization processes.

The basic philosophy followed in the book is that the availability of an *in situ* monitoring technique is essential for the successful implementation of a control strategy. The implementation, calibration, and testing of such a sensor is not straightforward and is therefore discussed in detail in this book. The control strategy is to a large extent dependent on the choice of the sensors, but it is also related to the crystallization system, the product specifications, and the available equipment. The other aspect emphasized in this book is the possibility of applying a model-based control strategy. This choice leads to flexible and cost-effective operations of continuously and batchwise operated industrial crystallizers. One of the key factors for such an approach is the availability of generalized rigorous process models which are easily tunable for the specific application and which can describe the evolution of the product quality in industrial crystallizers in a broad range of process conditions. This so-called master model, which in most cases is a complex nonlinear model, can then be used, provided that the appropriate tools are available, for optimization of the process conditions or trajectories, for a controllability analysis and after appropriate model reductions for the dynamic observer and the model predictive controller. This approach, which has successfully been developed recently, will be discussed in detail in this book. However, also more traditional single loop control strategies to improve the reproducibility and the product quality will be discussed extensively.

The first part of the book provides the reader with an overview on the state-of-the-art on instrumentation and methodologies for the online or *in situ* monitoring of relevant process variables in process environments, aiming at the online analysis and control of the crystallization process. After a first chapter where a number of different techniques for the characterization of the CSD are discussed and compared, instruments of commercial instruments, capable of determining either online or *in situ* some aspects of the CSD, are described (Chapters 2–5).

Traditional and new sensors for the measurement of nucleation and solubility points of solution are illustrated in Chapter 6. These techniques allow the determination of the metastable range width, which is the basis for the development of any crystallization process. Many measurements techniques are present in the literature, such as those based on dielectric constant (Hermanto *et al.*, 2011), calorimetric analysis (Lai *et al.*, 2011), and conductivity (Genceli, Himawan, and Witkamp, 2005), but those based on turbidity and ultrasound analysis are the most common and reliable ones, and moreover provided by relatively cheap instruments.

The online measurement of the solute concentration may be provided by a relatively cheap instrument, by an online refractometer, or by more sophisticated and expansive ones, as those based on ATR FTIR and Raman spectroscopy. These latter techniques, discussed in a recent paper of Kadam *et al.* (2010), are now currently adopted in investigation on pharmaceutical products, whereas that based on refractometry is advantageously applied in the sugar industry. All these techniques are widely reported in Chapters 7–10.

The second part covers the dynamic control of batch and continuously operated industrial crystallizers. In this part dynamic models suitable for model-based control strategies are discussed, as well as methods for parameter estimation and validation. Also the application of these models for the optimization of the process conditions is described. The basic, model free, control of batch operated crystallization processes is discussed in Chapters 11 and 12. In these Chapters particular attention is given to the control of the supersaturation profile during the batch process, based on a predefined recipe. The seeding technique is also examined and its optimal application is discussed with regard to the crystallization phenomena kinetics. This technique may have a very important role to control the initial phase of a batch process where the initial distribution of the crystals is generated or added to the crystallizer.

Advanced recipe and model based control is discussed in Chapters 13 and 14. In both process models are used to determine an optimal profile to achieve the desired process or product performances. In Chapter 13 the FB control is applied on the basis of an off line determined recipe or profile, while in Chapter 14, closed loop implementations of the model based control strategy are illustrated using state estimators and a real time optimization of the trajectory. This latter approach allows an early detection and feedback of process disturbances. The control of continuously operated crystallization processes is treated in Chapters 15 and 16. Firstly the main manipulation technique, that is, the one based on fines removal, is presented. Then in Chapter 16, the application of a model predictive control (MPC) for continuous crystallization processes is introduced. Both single loop control strategies as well as multivariable predictive control strategies are discussed. This chapter also gives a introduction into the principles, the design, and the implementation of MPC, including the necessary state estimation, are discussed in detail and some application examples are given.

Finally, in Chapter 17 the main types of crystallizers adopted in continuous crystallization process involving commodities are described together with their P&I schemes. This contribution is given by one of the leader worldwide companies in the design and construction of industrial crystallizers. The choice of the main instruments to be adopted for industrial units and their location inside the crystallizer, together with a discussion on the control valve to be used, completes information of the crystallizer's design.

Herman J. M. Kramer
Angelo Chianese

References

FDA, US Department of Health and Human Services (2004) Guidance for industry – PAT, http://www.fda.gov/Drugs/GuidanceComplianceRegulatoryInformation/Guidances/ucm064998.htm.

Genceli, F.E., Himawan, C., and Witkamp, G.J. (2005) Inline determination of supersaturation and metastable zone width of $MgSO_4 \cdot 12H_2O$ with conductivity and refractive index measurement techniques. *J. Cryst. Growth*, **275** (1–2), 1757–1762.

Hermanto, M.W., He, G.A, Tjahjono, M., Chow, P.S., Tan, R.B.H., and Garland, M. (2011) Calibration of dielectric constant measurements to improve the detection of cloud and clear points in solution crystallization. *Chem. Eng. Res. Des.*, in press j.cherd.2011.04.012.

Kadam, S.S., Van Der Windt, E., Daudey, P.J., and Kramer, H.J.M. (2010) A comparative study of ATR-FTIR and FT-NIR spectroscopy for *in-situ* concentration monitoring during batch cooling crystallization processes. *Cryst. Growth Des.*, **10** (6), 2629–2640.

Lai, X.A., Roberts, K.J.A., Svensson, J.B., and White, G. (2011) Reaction calorimetric analysis of batch cooling crystallization processes: studies of urea in supersaturated water–methanol solution. *Cryst. Eng. Commun.*, **13** (7), 2505–2510.

Nagy, Z.K., Fujiwara, M., and Braatz, R.D. (2008) Modelling and control of combined cooling and antisolvent crystallization processes. *J. Process Control*, **18**, 856–864.

Seki, H., Amano, S., and Emoto, G. (2010) Modeling and control system design of an industrial crystallizer train for para-xylene recovery. *J. Process Control*, **20**, 999–1008.

Sheikhzadeh, M., Trifkovic, M., and Rohani, S. (2008) Adaptive MIMO neuro-fuzzy logic control of a seeded and an unseeded anti-solvent semi-batch crystallizer. *Chem. Eng. Sci.*, **63** (5), 1261–1272.

Trifkovic, M., Sheikhzadeh, M., and Rohani, S. (2009) Multivariable real-time optimal control of a cooling and antisolvent semibatch crystallization process. *AIChE J.*, **55** (10), 2591–2602.

Valencia Peroni, C., Parisi, M., and Chianese, A. (2010) Hybrid modelling and self-learning system for dextrose crystallization process. *Chem. Eng. Res. Des.*, **88**, 1653–1658.

Woo, X.Y., Tan, R.B.H., and Braatz, R.D. (2011) Precise tailoring of the crystal size distribution by controlled growth and continuous seeding from impinging jet crystallizers. *Cryst. Eng. Commun.*, **13** (6), 2006–2014.

List of Contributors

Marco Bravi
Sapienza University of Rome
Department of Chemical
Engineering
Materials Environment
Via Eudossiana 18
00184 Rome
Italy

Angelo Chianese
Sapienza University of Rome
Department of Chemical
Engineering Materials
Environment
Via Eudossiana 18
00184 Rome
Italy

Jeroen Cornel
ETH Zurich
Institute of Process Engineering
Sonneggstrasse 3
8092 Zurich
Switzerland

Eugenio Fazio
University of Rome La Sapienza
Department of Fundamental and
Applied Sciences
Via Antonio Scarpa 16
00161 Rome
Italy

Adrie E.M. Huesman
Delft University of Technology
Delft Center for Systems and
Control
Mekelweg 2
2628 CD Delft
The Netherlands

Matthew J. Jones
AstraZeneca R&D
Global Medicines Development
Pharmaceutical Development
15185 Södertälje
Sweden

Somnath S. Kadam
Delft University of Technology
Process & Energy Laboratory
Leeghwaterstraat 44
2628 CA Delft
The Netherlands

Alex N. Kalbasenka
Delft University of Technology
Delft Center for Systems and
Control
Mekelweg 2
2628 CD Delft
The Netherlands

Michel Kempkes
ETH Zurich
Institute of Process Engineering
Sonneggstrasse 3
8092 Zurich
Switzerland

Herman J.M. Kramer
Delft University of Technology
Process & Energy Laboratory
Leeghwaterstraat 44
2628 CA Delft
The Netherlands

Christian Lindenberg
ETH Zurich
Institute of Process Engineering
Sonneggstrasse 3
8092 Zurich
Switzerland

Marco Mazzotti
ETH Zurich
Institute of Process Engineering
Sonneggstrasse 3
8092 Zurich
Switzerland

Klas Myréen
K-Patents Process Instruments
Elannontie 5
01510 Vantaa
Finland

Mariapaola Parisi
Sapienza University of Rome
Department of Chemical Engineering
Materials Environment
Via Eudossiana 18
00184 Rome
Italy

Franco Paroli
GEA Process Engineering S.p.A.
Centro Direzionale Milano 2
Palazzo Canova
20090 Segrate (MI)
Italy

Jochen Schöll
Mettler-Toledo AutoChem
Sonnenbergstrasse 74
8603 Schwerzenbach
Switzerland

Joachim Ulrich
Martin-Luther-Universität
Halle-Wittenberg
Zentrum für Ingenieurwissenschaften/TVT
06099 Halle (Saale)
Germany

1
Characterization of Crystal Size Distribution

Angelo Chianese

1.1
Introduction

Crystalline population coming out from a crystallizer is characterized by its size distribution, which can be expressed in different ways. The crystal size distribution ("CSD") may, in fact, be referred to the number of crystals, the volume or the mass of crystals with reference to a specific size range, or the cumulative values of number, volume or mass of crystals up to a fixed crystal size. The first approach refers to a density distribution, whereas the second one to a cumulative size distribution.

However, it is also useful to represent the CSD by means of a lumped parameter as an average size, the coefficient of variation, or other statistical parameters which may be adopted for the evaluation of a given commercial product.

In this section the more usual ways to represent both the whole CSD and the lumped CSD parameters are presented.

1.2
Particle Size Distribution

The particle size distribution may be referred to the density distribution or cumulative distribution. Each distribution may be expressed in number, volume, or mass of crystals.

The cumulative variable, $F(L)$, expresses number, volume, or mass of crystals per unit slurry volume between zero size and the size L, whereas the density distribution function, $f(L)$, refers to number, mass, or volume of crystals per unit slurry volume in a size range, whose average size is L.

The relationship between the cumulative size variable and the density distribution size one is as follows:

$$F(L) = \int_0^L f(L) \, dL \quad (1.1)$$

Industrial Crystallization Process Monitoring and Control, First Edition. Edited by
Angelo Chianese and Herman J. M. Kramer.
© 2012 Wiley-VCH Verlag GmbH & Co. KGaA. Published 2012 by Wiley-VCH Verlag GmbH & Co. KGaA.

1 Characterization of Crystal Size Distribution

Table 1.1 Cumulative and density variables.

Cumulative distribution			Density distribution		
Name	Symbol	Unit	Name	Symbol	Unit
Number	$N(L)$	#/m³ slurry	Population density	$n(L)$	#/m³ slurry m
Volume	$V(L)$	m³ crystal/m³ slurry	Volume density	$v(L)$	m³ slurry/m³ slurry m
Mass	$M(L)$	kg crystal/m³ slurry	Mass density	$m(L)$	kg crystal/m³ slurry m

or in the reverse form:

$$f(L) = \frac{dF(L)}{dL} \tag{1.2}$$

In Table 1.1 the expression of cumulative and density function variables referred to the number, volume, or mass of crystals is reported.

Examples of number and volume distributions are reported in Figure 1.1.

In spite of the three geometric dimensions of crystals, the CSD is usually referred to just one dimension, the so-called characteristic one, which is related

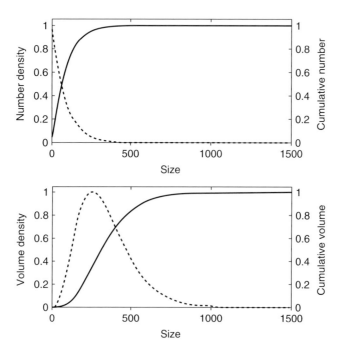

Figure 1.1 Cumulative and volume crystal size distributions (continuous line for the cumulative distribution and dotted lines for the density distribution).

to the adopted measurement technique. In the case of crystal size measurement by sieving the characteristic dimension is the second one, corresponding to the wire mesh length. Otherwise, if a laser diffraction-based analyzer is used, the characteristic dimension is the length given by the instrument, falling between the first and the second crystal dimension.

The most used density distribution variable is the crystal population density, $n(L)$. It can be used to estimate the total number, N_T, the total surface A_T, and total mass, M_T, of crystals by means of the following expressions:

$$N_T = \int_0^\infty n(L) dL \tag{1.3}$$

$$A_T = \int_0^L k_s(L) L^2 n(L) dL \tag{1.4}$$

$$M_T = \rho \int_0^L k_v(L) L^3 n(L) dL \tag{1.5}$$

where ρ is the crystal mass density and $k_s(L)$ and $k_v(L)$ are the surface shape factor and the volume shape factor of the crystals of size L, respectively, derived from the relationships

$$A(L) = k_s(L) L^2 \quad \text{and} \quad V(L) = k_v(L) L^3 \tag{1.6}$$

$A(L)$ and $V(L)$ are the surface and the volume of a single crystal, respectively. When the CSD is measured by sieving, the values of $k_v(L)$ can be easily determined, as follows:

$$k_v(L) = \frac{M_C(L)}{\rho L^3} \tag{1.7}$$

where $M_C(L)$ is the average mass of a single crystal retained over a sieve and L is the mean value between the dimension of two subsequent sieves, the considered and the upper ones. The surface shape factor is more difficult to calculate, since the average surface of the crystals has to be evaluated. The values of the shape factors depend on the crystal habit, which may qualitatively be defined as spherical, cubic, granular, needles, and so on. For a spherical particle the characteristic dimension is the diameter, the volume shape factor is equal to $\pi/6$, and the surface shape factor is π. Some examples of shape factors are reported in Table 1.2 (Mersmann, 2001).

In the case of size measurement by sieving, it is possible to convert the sieving data to crystal population density by means of the equation

$$n(L) = \frac{M(L)}{k_v \rho L^3 \Delta L} \tag{1.8}$$

where is $M(L)$ the overall crystal mass between two subsequent sieves, L and ΔL are the mean size and the size difference of the two subsequent sieves opening, respectively.

1 Characterization of Crystal Size Distribution

Table 1.2 Examples of shape factors.

Geometric shape	k_v	k_s
Sphere	0.524	3.142
Tetrahedron	0.118	1.732
Octahedron	0.471	3.464
Hexagonal prism	0.867	5.384
Cube	1.000	6.000
Needle 5 × 1 × 1	0.040	0.880
Plate 10 × 10 × 1	0.010	2.400

1.3
Particle Size Distribution Moments

The *i*th moment of a size distribution is defined as

$$m_i = \int_0^\infty L^i n(L) \mathrm{d}L \tag{1.9}$$

The moments are useful to calculate the mean size, as follows:

$$\bar{L}_{i+1,i} = m_{i+1}/m_i \tag{1.10}$$

According to the value of *i* between 0 and 3 used, the mean size may be referred to the number of crystals, their length, surface, and volume, respectively. The most popular expressions of mean size, derived from the moments ratio, are the number-based and the volume-based mean size (Randolph and Larson, 1988), that is,

$$\bar{L}_{1,0} = \int_0^\infty L n(L) \mathrm{d}L \bigg/ \int_0^\infty n(L) \mathrm{d}L \tag{1.11}$$

$$\bar{L}_{4,3} = \int_0^\infty L^4 n(L) \mathrm{d}L \bigg/ \int_0^\infty L^3 n(L) \mathrm{d}L \tag{1.12}$$

The number, the length, the surface area, and the volume of the particles are directly related to the moments of the distribution. The relevant expression, when the shape factor is constant with size, is reported in Table 1.3.

The coefficient of variation (CV) represents the spread of the size distribution around the mean size. It is the ratio between the distribution standard deviation and the mean size.

Often, the median size, L_{median}, instead of the mean size is used. It is the mean abscissa of a graph of cumulative volume/mass fraction versus size. Note that in general the median size is different from the mean size.

Table 1.3 Properties of distribution based on the moments.

Property	Definition	Dimension
Total number	$N_T = m_0$	(#/m³)
Total length	$L_T = m_1$	(m/m³)
Total surface area	$A_T = k_s m_2$	(m²/m³)
Total volume	$V_T = k_v m_3$	(m³/m³)
Mean (number based)	$L_{1,0} = m_1/m_0$	(m)
Mean (volume based)	$L_{4,3} = m_4/m_3$	(m)
Coefficient of variation (volume based)	$CV = \sqrt{\dfrac{m_3 m_5}{m_4^2} - 1}$	–

1.4 Particle Size Distribution Characterization on the Basis of Mass Distribution

The calculation of characteristic sizes of the CSD does not necessarily requires to pass through the calculations of values of the crystals population density. The most easy way is, in fact, to use directly the mass fraction of crystals measured by sieving. The mass fraction of crystals of an average size equal to L can be calculated as

$$x(L) = \frac{M(L)}{M_T} \tag{1.13}$$

Often the mass fraction is represented by means of the histogram reported in Figure 1.2, which is called the *frequency histogram*.

Sequentially adding each segment of the frequency diagram gives the cumulative distribution in terms of the mass fraction.

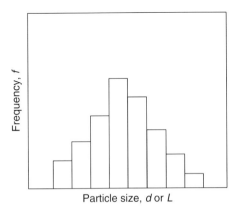

Figure 1.2 Frequency histogram.

The size distribution may be characterized by two parameters: the mass mean size L_{wm} and the coefficient of variation, that is,

$$L_{wm} = \sum_i x_i(L) L \tag{1.14}$$

and

$$CV = \frac{1}{L_{wm}} \sqrt{\frac{\sum_i (L_i - L_{wm})^2}{N - 1}} \tag{1.15}$$

References

Mersmann, A. (2001) *Crystallization Technology Handbook*, 2nd edn, Marcel Dekker, New York.

Randolph, A.D. and Larson, M.A. (1988) *Theory of Particulate Processes*, 2nd edn, Academic Press, San Diego, CA.

2
Forward Light Scattering
Herman J.M. Kramer

2.1
Introduction

Instruments based on laser diffraction to determine the particle size distribution (PSD) have been developed in the last few decades. The first instruments such as the legendary 2600 series particle sizers of Malvern Instruments (UK) measured the laser diffraction pattern with a series of semiconcentric ring detectors to record the low angle diffraction pattern in the forward direction. The measured axisymmetric diffraction pattern of all the rings represents then the signature of the crystal size distribution of the suspension. Thanks to Switchenbank *et al.* (1977), this diffraction pattern can be deconvoluted into the PSD using a inversion technique based on Fraunhofer diffraction. Over the past few years laser diffraction instruments have been modified and improved. For instance, the latest versions have detectors that are not semiconcentric anymore, have much smaller dimension, and cover an angular sector only with a surface area that follows a logarithm progression from the center to the outside. Also the measurement of the wide angle and backward scattering, the use of blue lasers, and the implementation of the more complete Mie scattering theory to describe the laser diffraction have contributed to extend the particle size range which can be covered in a single measurement.

Laser diffraction is applied widely and has become the most used and the standard sizing technique. The reasons for this success of these instruments are that they are easy to operate, offer a nondestructive way for fast measurement of the crystal size distribution, and produce reproducible results without the need for extensive calibrations procedures.

The major drawback of forward light scattering instruments is that a low particle concentration is required to avoid multiple scattering effects, so mostly industrial crystal suspensions must be diluted before they can be analyzed by this method. This requirement limits the application of these instruments *in situ* in a process. An additional problem forms the sensitivity of the instruments for the shape of the particles. The instruments measure a projected area of the particles in a 2D plane, which in the case of nonspherical particles is dependent on the shape and the orientation of the particles in the measurement cell. A spherical equivalent diameter is then used for the sizing of the particles which gives rise to deviations from the true particle size (Neumann and Kramer, 2002; Erdoğan et al., 2010).

2.2
Principles of Laser Diffraction

A parallel beam from a coherent monochromatic light source falling on an ensemble of particles creates a composite scatter pattern. This scatter pattern, which is a function of the size and the optical properties of the particles, can be measured with a lens and a spatial detector placed in the focal plane of the lens (see Figure 2.1). An intensity pattern as a function of the scatter angle with respect to the optical axis of the instruments can then be recorded. At an angle of $0°$ the undeflected light is gathered giving rise to a very intense spot. In most cases a hole is made in the detector to prevent the blinding of the detector elements in the neighborhood of this location. This so-called obscuration signal is measured with a separate detector and forms a measure of the particle concentration in the measurement volume. As the size information in the low angle scatter pattern is rather limited for the smaller particles, the modern instruments measure besides the low scatter angle pattern the wide angle and the backscatter intensities (see Figure 2.2). The low angle scatter is determined by the focal plane detectors, placed perpendicular on the

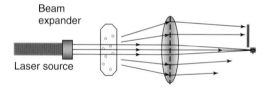

Figure 2.1 Schematic view of a laser diffraction instrument measuring the low angle scattering only.

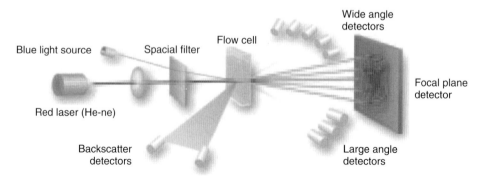

Figure 2.2 Schematic view of a modern FLS instrument.

incident beam (see Figure 2.2), while separate wide angle and backscatter detectors are present to increase the sensitivity of the instruments toward the low size range. The size resolution of the instrument is in principle limited by the wavelength of the used laser light. However equipped with this wide angle detector and dual laser (blue and red) size resolutions between 50 and 100 nm and several millimeters are claimed.

The scatter pattern as measured with a number of detector elements is the result of the scatter signature of all the particles present in the measurement volume. This can be described as follows:

$$\begin{pmatrix} L_1 \\ \cdot \\ \cdot \\ L_n \end{pmatrix} = \begin{pmatrix} A_{1,1} & \cdot & \cdot & A_{1,m} \\ \cdot & \cdot & \cdot & \cdot \\ \cdot & \cdot & \cdot & \cdot \\ A_{n,1} & \cdot & \cdot & A_{n,m} \end{pmatrix} \begin{pmatrix} Q_1 \\ \cdot \\ \cdot \\ Q_m \end{pmatrix} \quad (2.1)$$

where
L = measured light energy vector
A = scatter matrix
Q = unknown size distribution
n = number of detector elements
m = number of size classes
$A_{i,j}$ = element of the scatter matrix A, representing the contribution of the total scatter pattern for a particle of class i on the detector element j, computed according to the light scatter theory used.

This equation is valid only when the scatter pattern of the individual particles is not influenced by the presence of the other ones. This means that multiple scattering must be avoided, which limits the particle concentration to about 1% in volume.

To deduce the size distribution from a measured scatter pattern, the scatter matrix A must be known. In addition a deconvolution technique must be used to solve the unknown size distribution from the measured light energy vector L using Equation (2.1). These aspects will be discussed in the next sections.

In Figure 2.3, the scatter pattern of a number of mono disperse size distributions is given. In this example normal distributions are chosen around 50, 100, 150, 200, and 250 μm with a standard deviation of 10 μm. The scatter pattern is calculated for a Malvern Master Sizer X equipped with a 100 mm lens. It is clear that the larger particles scatter at smaller angles, while the smaller particles show a much broader signature. In Figure 2.4, the scatter signature of the mixture is given, which is obtained by mixing the five samples at equal volume fractions. This clearly illustrates the difficult task of the diffraction instrument. In spite of the clearly separated sizes of the five samples, the scatter pattern is rather smooth and it will be a difficult task to find a unique solution for the composition of the sample.

Note that the diffraction patterns shown in Figures 2.3 and 2.4 are determined by the diffraction pattern of the particles and the geometry and sensitivity of the diffraction instrument.

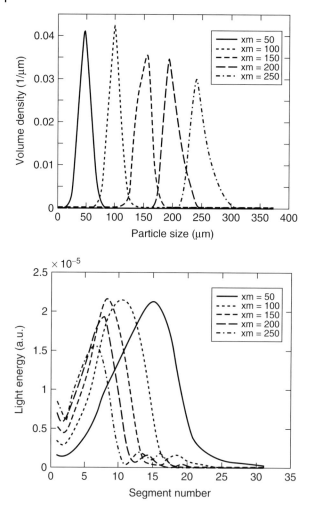

Figure 2.3 Volume density plots of five monodisperse suspensions and their scatter patterns on a Malvern Master Sizer X equipped with a 100 mm lens. Higher segment numbers mean larger scatter angles.

2.3
Scatter Theory

The light scatter theory describes the interaction of the electromagnetic field of the incident light with the particles present in the suspension. The following phenomena can be mentioned (see Figure 2.5): reflection on the surface of the particle and inside the particle, and refraction from the medium into the particle and from the particle into the medium, diffraction, and absorption. The relative importance of these phenomena is determined by the properties of the particles

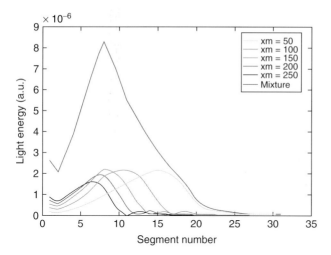

Figure 2.4 Scatter pattern of five samples and of their mixture.

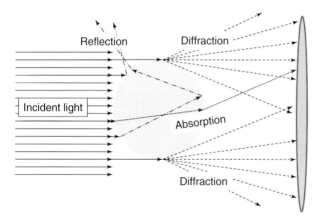

Figure 2.5 Scatter phenomena around a particle.

such as size, refractive index, homogeneity, and the wavelength of the incident light beam. A number of scatter theories have been developed for specific applications.

A brief discussion will be given here of the most used theories, which describe part of the above-mentioned phenomena under the following assumptions:

- The particles are homogeneous and spherical.
- Single scattering: only one scatter event occurs before the light falls on the detector.
- Incoherent scattering: the particles are randomly oriented in the measurement volume.
- Independent scattering: the scatter pattern of a particle is not influenced by the presence of the other particles in the measurement volume.

These assumptions limit the application of the technique to low particle concentrations (in general below 1% by volume). From the last two assumptions it can be deduced that the measured scatter pattern of the ensemble of particles is a linear combination of the scatter pattern of the individual particles.

2.3.1
Generalized Lorenz–Mie Theory

The rigorous description of the light scattering is given by the generalized Lorenz–Mie theory (Bohren and Huffman, 1983; Boxman, 1992). It uses the Maxwell equations at every point in space and can be solved for small scatter angles and spherical particles (Boxman, 1992), although the calculation time increases rapidly for larger values of the dimensionless particle size α:

$$\alpha = \frac{\pi}{\lambda} x \tag{2.2}$$

with λ the wavelength of the light (m) and x the equivalent circular area diameter. For different particle shapes and large particles, the solution of the Lorenz Mie becomes unreliable and in those cases the scatter pattern can more efficiently be calculated by approximate theories. The best known are

- anomalous diffraction (Van der Hulst, 1957);
- Fraunhofer diffraction.

All three theories have a limited validity due to additional assumptions made to simplify the description of the diffraction pattern.

2.3.2
Anomalous Diffraction

The anomalous diffraction theory (Van der Hulst, 1957) is valid only for particles much larger than the wavelength of the laser beam and for values around 1 for the relative refractive index m, which is defined as

$$m = \frac{m_{par}}{m_{med}}$$

in which m_{par} and m_{med} are the complex refractive indices for the particles and the medium, respectively. In this theory the assumption is made that the diffraction can be described by a combination of geometrical optics and diffraction. The light, which is refracted through the large particles, obtains a considerable phase shift with respect to the undisturbed light due the difference in refractive index between the particle and the medium. This phase shift gives rise to a distortion of the diffraction and is taken into account in the calculation of the diffraction signature. In the case of absorbing particles, the theory is no longer valid.

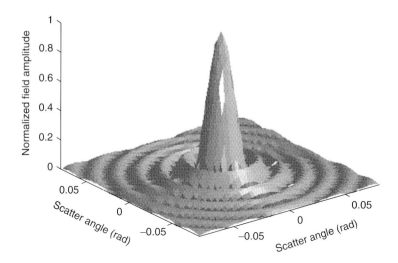

Figure 2.6 Diffraction pattern of spherical particles in the Fraunhofer regime; $A = 200$.

2.3.3
Fraunhofer Diffraction

In this theory the contributions of reflection and refraction are neglected completely. It is valid only if the following relation holds (Van der Hulst, 1957):

$$\alpha = \frac{3}{|m-1|}$$

For sufficiently large particles the scatter signature is accurately described by this scalar diffraction theory. For spherical particles, the angular irradiance is described by

$$I(\theta) = I_0 \left[\frac{J_1(\alpha \sin(\theta))}{\alpha \sin(\theta)} \right]^2$$

with J_1 the Bessel function of the first kind, θ the observation angle, and I_0 the incident light intensity on the particle.

The diffraction pattern of a spherical particle is the well-known Airy pattern as shown in Figure 2.6.

2.4
Deconvolution

When the size distribution is known, the prediction of the measured light energy using a scatter matrix is straightforward using the equation

$$L = AQ \tag{2.3}$$

A laser diffraction instrument however has to deal with the inverse problem. The unknown size distribution has to be deconvoluted from the measured light energy vector. This inversion problem appears to be a difficult one because the scatter matrix is ill-conditioned, due to the steepness of the diffraction function. There are a number of methods to solve this inversion problem. The most used ones, that will be discussed shortly below are:

- direct inversion with a nonnegativity constraint;
- Philips Twomey inversion;
- iterative methods.

2.4.1
Direct Inversion Using the Nonnegativity Constraint

Direct inversion of matrix A gives the solution $Q = A^{-1} L$ provided that A is a square matrix. In case A is not a square, the solution becomes

$$Q = (A^T A)^{-1} A^T L \tag{2.4}$$

Solving this equation generally leads to negative values in the size vector Q which have physically no meaning. The desired solution must therefore be a least-squares solution with a nonnegativity constraint applied to the solution vector (Boxman, 1992).

2.4.2
Philips Twomey Inversion Method

This method is often used (Boxman, 1992; Riebel and Löfffler, 1989), especially when the set of linear equations contains many variables and is highly unstable. The method solves the set of equations by smoothing the size distribution vector, adjustable with a smoothing factor γ. The method was developed by Philips (1962) and later improved by Twomey (1977). The method basically leads to a least-squares solution with a variable amount of smoothing. The size vector becomes

$$Q = (A^T A + \gamma K^T K)^{-1} A^T L \tag{2.5}$$

with K the smoothing matrix,

$$K = \begin{pmatrix} 0 & 0 & 0 & 0 & . & 0 & 0 & 0 & 0 \\ -1 & 2 & -1 & 0 & . & 0 & 0 & 0 & 0 \\ 0 & -1 & 2 & -1 & . & 0 & 0 & 0 & 0 \\ . & . & . & . & . & . & . & . & . \\ . & . & . & . & . & . & . & . & . \\ . & . & . & . & . & . & . & . & . \\ 0 & 0 & 0 & 0 & . & -1 & 2 & -1 & 0 \\ 0 & 0 & 0 & 0 & . & 0 & -1 & 2 & -1 \\ 0 & 0 & 0 & 0 & . & 0 & 0 & 0 & 0 \end{pmatrix} \tag{2.6}$$

which is a square matrix with column size of Q. Matrix K relates the different size classes to each other. The main advantage of this method is the relative small calculations times needed on a computer. The main disadvantage is that there is no systematic way to find an optimal value for γ, which depends on the absolute values of the matrix elements. Secondly, the size distributions are broadened. In spite of the visually more attractive smooth PSD, the residual error always becomes larger when smoothing is applied.

2.4.3
Iterative Methods

Iterative methods minimize the value of the error function E defined by

$$E(Q) = L - AQ = L_{\text{meas}} - L_{\text{calc}} \qquad (2.7)$$

The minimization is undertaken by optimizing the size vector Q. Several iteration schemes can be used (see Weiss et al., 1993).

2.5
The Effects of Shape

The shape of the crystals has a marked effect on its diffraction spectrum as illustrated in Figures 2.6 and 2.7, which show the diffraction patterns of a spherical- and rectangular-shaped particles, respectively. It will be clear that the particle shape has a considerable effect on the PSD measurements, as in commercial instruments the shape of the particles is not taken into account.

In a diffraction instrument, the diffraction pattern is measured of an ensemble of particles, which have in principle a random orientation. The resulting pattern will therefore be the average of the 2D projection of all possible orientations of the particles. This results in an apparent crystal size distribution (CSD) that is much broader than the real one and especially with elongated particles, the CSD gets

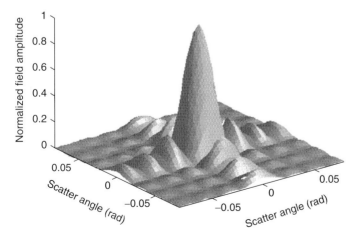

Figure 2.7 Diffraction pattern of a rectangular-shaped particle.

a multimodal character. To correct for these shape effects two strategies can be followed.

1) Calibration of the instrument by comparison of the measured size distribution with some independent size measurement device. This method is a practical approach and can be realized without involvement of the manufacturer. The fundamental problems like the broadening of the diffraction peaks are not solved however.
2) Recalculation of the scatter matrix based on information of the particle shape. To do this however one must be able to recalculate the scatter matrix used in the software and thus involves help of the manufacturer of the size instruments.

In the past attempts have been reported to determine the shape of particles using the azimuthal variations in the detector plane (Heffels et al., 1994). Unfortunately, at the moment no commercial instruments are available that use this technique.

2.6
Multiple Scattering

In industrial suspensions the particle concentration is often too high to avoid multiple scattering to occur. Therefore some of the commercial instruments include a patented multiple scattering algorithm. According to Hirleman (1988, 1990), multiple scattering of the laser beam can occur in two situations: (i) when the interparticle spacing is so small that the scattering characteristics of a particle depend on the positions and sizes of adjacent particles, (ii) when the optical path (extending in the direction of the incident radiation) is so large that a significant number of photons are scattered more than once before leaving the medium and reaching the detector. As a consequence, the diffraction angles are increased and the initial mathematical inversion procedure, based on the assumption of individual photon scattering of single particles, overestimated the small-particle population, and underestimated the width of the distribution.

To remedy this problem, Hirleman (1988, 1990) developed multiple scattering models. Combining the method of successive orders with the discrete ordinates approach similar to the one of Felton et al. (1985). This formulation assumes isolated particle light scattering, an axisymmetric diffraction pattern, particle characteristics (concentration and particle-size distribution) uniformly distributed in space, particles larger than the light wavelength and considers the scattered light in the near-forward direction only. Furthermore, the model was developed for a laser diffraction instrument with a specific circular diode series arrangement. The correction algorithm implemented in the Malvern instruments has been derived from Hirleman's model (Harvill, Hoog, and Holve, 1995) and allows measurements with transmission as low as 5% to be performed.

2.7
Application of Laser Diffraction for Monitoring and Control of Industrial Crystallization Processes

Application of laser diffraction to monitor the evolution of the CSD in industrial crystallization processes is rare and in industrial practice monitoring of the CSD is mostly realized by taking (dry) samples from the crystallizer followed by a CSD analysis in the laboratory (coulter counter, sieving, laser diffraction, etc.). Application of laser diffraction instruments directly in the process has a number of disadvantages when used in industrial environment. The most critical one is related to the fact that the use of this technique is restricted to the solid concentration of approximately 1 vol%, whereas the product concentration in industrial crystallizers is in generally an order of magnitude higher. In-line CSD monitoring therefore requires an on-line dilution system, decreasing the crystal concentration. The only alternative is the use of a multiple scattering algorithm as outlined in Section 2.6, which permits CSD monitoring of particle concentrations up to 4–5 vol%. However, the reconstructed CSD using a multiple scattering algorithm is of much lower quality, while the maximal usable particle concentration is not high enough for most industrial crystallization applications.

Automatic dilution systems have been reported to be capable of enabling real-time monitoring of the CSD in crystallization processes (Eek, 1995; Boxman, 1992; Neumann and Kramer, 2002), which enable the accurate CSD monitoring in real time in a crystallizer during several days using a laser diffraction instrument. The dilution systems only require a constant stream of dilution liquid and two programmable pneumatic valves to enable alternative background and sample measurements (see Figure 2.8). It has been shown that a laser diffraction instrument equipped with an automatic dilution system is quite capable of following the development of the CSD in an industrial crystallizer and that detailed information

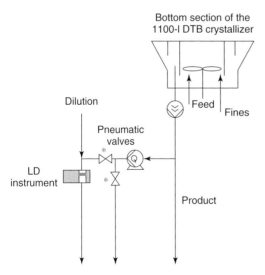

Figure 2.8 Dilution system for a real-time CSD measurement using laser diffraction.

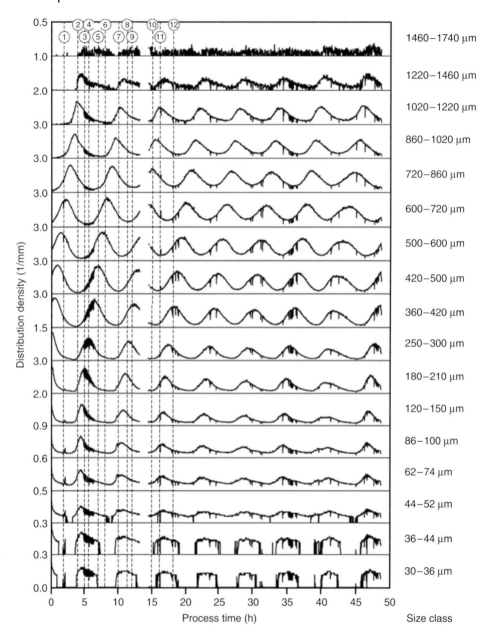

Figure 2.9 Evolution of the CSD in time for different size fractions in an evaporative DTB crystallizer, measured on-line using an automatic dilution system, and a Helos laser diffraction instrument. (From Neumann (2004).)

that is not restricted to the mean crystal size is obtained about the evolution of the CSD. This is demonstrated in Figure 2.9 from Neumann (2004), which shows the trend in the population density of 17 size fractions over a period of 45 h in an evaporative draft tube baffled (DTB) crystallizer of 1100 l. The CSD was measured using a HELOS/VARIO laser diffraction instrument after dilution of the crystal suspension with the saturated solution. The dilution ratio could be varied but was about 15 at steady state. The results shown in Figure 2.9 show a large detail that is obtained by the application of this measurement system. Peaks in the distribution, first seen in the small size fractions, are slowly moving to the larger size classes, clearly visualizing the dynamics of the growth of the crystals. Figure 2.9 also clearly shows that the crystallizer is operated in a limit cycle. This information would be of great value for the operators of the process to stabilize it and to direct the CSD toward a desired distribution. It is also shown that the systems are reasonably robust; in the 45 h operation time only one major blockage occurred after about 13 h which was solved by a rinsing the system.

Unfortunately the use of laser diffraction in combination with an automated dilution system is hardly applied yet. Apparently the fear that the system is not robust enough has prevented the general acceptance so far, despite that automated dilution systems are available commercially.

2.8 Conclusions

Forward light scattering is a powerful particle sizing technique, which is widely accepted as the standard technique. The following characteristic properties can be identified:

- Total measuring range 0.05–6000 µm.
- Sensor generates a projected area-based volume distribution.
- Resolution: high.
- Suitable for online applications (with an automatic dilution system).
- Sampling and dilution is needed down to a volume concentration of about maximal 1% in volume. Multiple scatter algorithms are available up to 4–5%.
- Calibration not needed.
- Shape effects complicate the size analysis.

It can be concluded that laser diffraction is a powerful measurement technique for the characterization of the CSD in crystallization processes. In combination with an automated dilution system online monitoring of the CSD is feasible and delivers valuable information on the dynamics of the process which can be used directly to control the crystal quality. Another important drawback is that the instrument is not capable of taking into account the shape of the crystals in a proper way. This is especially important for needle-like crystals, which are often encountered in organic or pharmaceutical crystallization systems.

References

Bohren, C.F. and Huffman, D.R. (1983) *Absorption and Scattering of Light by Small Particles*, John Wiley & Sons, Inc., New York.

Boxman, A. (1992) Particle size measurement for the control of industrial crystallisers. PhD thesis. Delft University of Technology, Delft, The Netherlands.

Eek, R. (1995) Control and dynamic modelling of industrial suspension crystallisers. PhD thesis. Delft University of Technology, Delft, The Netherlands.

Erdoğan, S.T., Nie, X., Stutzman, P.E., and Garboczi, E.J. (2010) Micrometer-scale 3-D shape characterization of eight cements: particle shape and cement chemistry, and the effect of particle shape on laser diffraction particle size measurement. *Cement Concrete Res.*, **40**, 731–739.

Felton, P.G., Hamidi, A.A., and Aigal, A.K. (1985) Measurement of drop size distribution in dense sprays by laser diffraction. *Proceedings of the 3rd International Conference Liquid Atomization and Spray Systems*, Institute of Energy, London.

Harvill, T.L., Hoog, J.H., and Holve, D.J. (1995) In-process particle size distribution measurements and control. *Part. Part. Syst. Charact.*, **12**, 309–313.

Heffels, C.M.G., Heitzmann, D., Hirleman, E.D., and Scarlett, B. (1994) The use of azimuthal intensity variations in diffraction patterns for particle shape characterisation. *Part. Part. Syst. Charact.*, **11**, 194–199.

Hirleman, E.D. (1988) Modelling of multiple scattering effects in Fraunhofer diffraction particle size analysis. *Part. Part. Syst. Charact.*, **5**, 57–65.

Hirleman, E.D. (1990) A general solution to the inverse near-forward scattering particle sizing problem in multiple scattering environment: theory. Second International Congress on Optical Particle Sizing, Tempe, AZ, pp. 159–168.

Neumann, A.M. (2004) Characterizing Industrial crystallizers of different scale and type. PhD thesis. Delft University of Technology, Delft, The Netherlands.

Neumann, A.M. and Kramer, H.J.M. (2002) A comparative study of various size distribution measurement systems. *Part. Part. Syst. Charact.*, **19**, 17–27.

Philips, B.L. (1962) A technique for numerical solution of certain integral equations of the first kind. *J. Assoc. Comput. Mach.*, **9**, 84–97.

Riebel, U. and Löffler, F. (1989) The fundamentals of particle size analysis by means of ultrasonic spectrometry. *Part. Part. Syst. Charact.*, **6**, 135–143.

Swithenbank, J., Beer, J.M., Taylor, D.S., Abbot, D., and McGreath, C.G. (1977) A laser diagnostic technique for the measurement of droplet and particle size distribution, *Experimental Diganostics in Gas Phase Combuslion Systems, Progress in Astronautics and Aeronautics*, vol. **53** (ed. B.T. Zinn), AIAA, New York.

Twomey, S. (1977) *Introduction into the Mathematics of Inversion in Remote Sensing and Indirect Measurements*, Elsevier, Amsterdam.

Van der Hulst, H.C. (1957) *Light Scattering by Small Particles*, Dover Publications, New York.

Weiss, M., Schmidt, M., and Bottlinger, M. (1993) Model based on line analysis of disperse systems on the example of digital Fourier spectroscopy. *Part. Part. Syst. Charact.*, **10** (6), 339–346.

Further Reading

Allen, T. (1990) *Particle Size Measurement*, Chapman & Hall, London, ISBN: 0 412 35070 X.

Dumouchel, C., Yongyingsakthavorn, P., and Cousin, J. (2009) Light multiple scattering correction of laser-diffraction spray drop-size distribution measurements. *Int. J. Multiphase Flow*, **35**, 277–287.

3
Focused Beam Reflectance Measurement

Jochen Schöll, Michel Kempkes, and Marco Mazzotti

3.1
Measurement Principle

In the focused beam reflectance measurement (FBRM), a solid-state laser light source produces a continuous beam of monochromatic light that is focused to a small spot at a constant distance on the surface of the probe window. A pneumatic or electrical motor is used to rotate the optics, such that the rotating, focused beam of laser light is constantly scanning over particles that are passing in front of the probe as shown in Figure 3.1. The suspended particles backscatter the laser light to the probe where the reflected light is detected. From the duration of the backscatter and the rotation velocity of the optics, the distance the beam has scanned over the particle surface, the so-called chord length, can be calculated. The resulting measurement is the chord length distribution (CLD).

Clearly, the measured CLD is a function of the number, dimension, and shape of the particles in the suspension. Due to the random orientation of the suspended particles and the random location where the beam can scan each of these particles, the particle size distribution (PSD) cannot be directly extracted from the CLD. Although many research efforts have been directed at the determination of the PSD from measured CLD (Wynn, 2003; Li and Wilkinson, 2005a,b; Ruf, Worlitschek, and Mazzotti, 2000; Worlitschek, Hocker, and Mazzotti, 2005; Kempkes, Eggers, and Mazzotti, 2008; Yu, Chow, and Tan, 2008), so far no generally applicable solution has been proposed. For industrial purposes it is therefore recommended to use the real-time CLD data directly as "fingerprint" of the process which is highly sensitive to changes in number and particle dimension, instead of extracting an accurate PSD out of the CLD data.

3.2
Application Examples

3.2.1
Solubility and Metastable Zone Width (MSZW)

A basic FBRM application is the automated determination of two fundamental parameters of a crystallization process, namely solubility information and metastable

3 Focused Beam Reflectance Measurement

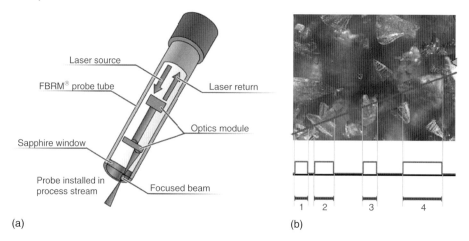

Figure 3.1 (a) Scheme of the FBRM probe. Particles in suspension backscatter the laser light emitted by the probe. (b) The number of particles and their scanned dimensions are recorded as the chord length distribution.

zone width (MSZW) determination, in combination with an automated lab reactor system (Barrett and Glennon, 2002). Although a simple turbidity probe can be used as well for such a task, the CLD data yield additional information regarding the relative nucleation and growth kinetics of the studied system, which is not revealed by the turbidity data.

3.2.2
Seed Effectiveness

FBRM is widely used to study the efficiency of a seeding event and to quantify the effectiveness of the seed material (Aamir, Nagy, and Rielly, 2010). Without such an *in situ* analytical tool it is difficult to evaluate seed effectiveness before the process end. One example has been presented by Lafferrère and coworkers who used FBRM together with an *in situ* microscope to study the impact of a seeding protocol on crystal growth and primary and secondary nucleation (Lafferrère, Hoff, and Veesler, 2004). The investigated system exhibited under certain process conditions a liquid–liquid phase separation, also called *"oiling-out"* and the *in situ* analytical tools were used to characterize the phase separation region in the phase diagram. Via controlled seed addition in the metastable zone the oiling-out could be prevented and a reproducible process was obtained.

3.2.3
Polymorph Transformations

Different polymorphic forms of a given molecule often exhibit a significantly different crystal shape. In such cases optical *in situ* measurement techniques such

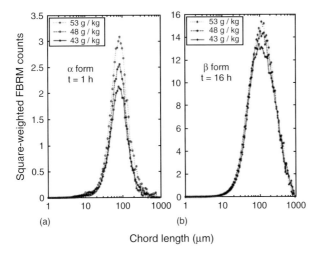

Figure 3.2 Experimental chord length distributions of the solvent mediated transformation process of L-glutamic acid at 45°C at three different initial concentrations and two different batch times.

as FBRM or inline microscopy can be used to monitor a polymorph transformation that can occur during the course of a crystallization process, although such optical techniques do not contain any information about the structure of the crystal lattice (Schoell *et al.*, 2006a).

Figure 3.2a represents the CLDs recorded 1 h after the start of the process and shows the distribution of the α polymorph. The CLDs reported on Figure 3.2b have been recorded after 16 h at the end of the transformation process, and show the final CLD of the β polymorph. The effect of an increasing solid phase concentration due to higher concentration levels can be directly observed in the chord length data. However, the polymorph transformation was monitored in combination with *in situ* microscopy, ATR-FTIR, and Raman spectroscopy in this case (Schoell *et al.*, 2006a).

The sensitivity of FBRM to nucleation events has also been used to understand the crystallization of diastereomers, permitting the identification and minimization of secondary nucleation, and consequently, minimization of the undesired diastereomer, thus increasing product quality and reducing cycle times (Mousaw, Saranteas, and Prytko, 2008).

3.2.4
Effect of Different Impurity Levels

Changing raw materials for a crystallization process can result in varying levels of impurities, which in turn may dramatically affect the thermodynamics, as well as the crystal growth and nucleation kinetics in the system. Black and coworkers used FBRM to study the impact of specific impurities on crystallization kinetics (Scott and Black, 2005). They observed that crystallization using pure materials resulted in the attainment of an endpoint very quickly, whereas the presence of impurities slowed down crystal growth kinetics; thus, the endpoint was reached much later. FBRM was used to monitor the relative growth and nucleation rates at different impurity levels and allowed the determination of the impact of these impurities on cycle times and yield.

3.2.5
Nucleation Kinetics

Besides the relative determination of particle formation kinetics, FBRM has been used to determine nucleation kinetics in a first-principles approach, by accurate induction time measurements as a function of supersaturation. The combination of ATR-FTIR spectroscopy and FBRM allowed for precise determination of the induction time, that is, the time span between attainment of a homogeneous supersaturation level throughout the reactor and the detection of particle formation (Figure 3.3) (Schoell et al., 2006b).

3.2.6
Improved Downstream Processing

Average particle size, particle shape, and size distribution width have a significant impact on the different unit operations in downstream processing, that is, filtration, washing, and drying. Kim and coworkers used FBRM to optimize a crystallization

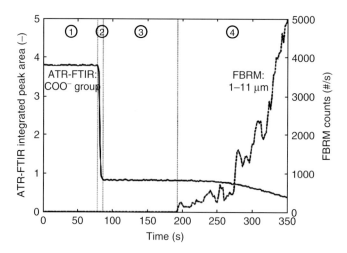

Figure 3.3 Time evolution of the ATR-FTIR and FBRM signal at a supersaturation of $S = 5.0$. The four phases of the experiment (1 lag time, 2 establishment of supersaturation, 3 induction period, and 4 detectable particle formation) can be clearly distinguished (Schoell et al., 2006b).

process with the objective of minimizing filtration and drying times and the facilitation of powder handling (Kim *et al.*, 2005). In this particular example, FBRM was used to characterize the effect of an acid addition profile and the seed mass on the particulate product. Moreover, FBRM was used to design an optimized drying process to ensure consistent dissolution kinetics of the final drug product.

A similar application of the FBRM as a "soft sensor" is also reported, where, for a particular crystallization process optimization, the CLD is related directly to functional properties of the crystalline product, for example, filterability (Togkalidou *et al.*, 2001). In these cases, one uses the median of the CLD (with no weighting) as an indication of the amount of fines present, which are in turn directly related to the filterability.

3.2.7
Process Control

Despite several decades of crystallization research, batch crystallization process control is still mainly based on indirect parameters such as residence times, temperature, and antisolvent/cooling profiles. The advent of *in situ* process analytical

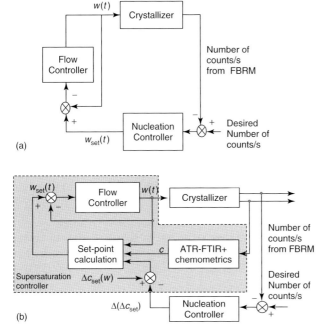

Figure 3.4 Block diagrams of (a) the DNC approach, where $w(t)$ is the actual antisolvent/solvent flowrate into the crystallizer; $w_{set}(t)$ is the antisolvent/solvent flowrate set-point given by the controller to the pump (Abu Bakar *et al.*, 2009). (b) Schematic diagram of the adaptive concentration control combining liquid phase (via ATR-FTIR) and solid phase characterization (using FBRM) (Woo *et al.*, 2009).

tools for the liquid and solid phases has the potential to open the door to closed-loop process control.

The implementation of the control loop can be based either on the supersaturation level of the liquid phase (Liotta and Sabesan, 2004), the direct control of the solid phase (Abu Bakar et al., 2009), or a combination of both (Woo et al., 2009). Figure 3.4 shows schematic diagrams of the direct nucleation control approach based on the FBRM characterization of the solid phase (Abu Bakar et al., 2009) and the combination of liquid- and solid phase-based control (Woo et al., 2009).

The main advantage of combining the characterization of liquid and solid phase, that is, quantifying the supersaturation in the liquid phase using ATR-FTIR and the particulate product via FBRM, for the control approach presented in Woo et al. (2009) is a lower sensitivity to process disturbances, changing thermodynamics, or kinetics with respect to the traditional fixed batch recipe approach. A second advantage with respect to process control schemes based on first principle kinetics is that minimal *a priori* information is required and a time-consuming kinetics determination is not needed. Although such a control approach might produce increased variation of process parameters, it has potential to minimize variations of important product quality attributes, that is, particle size, size distribution, and purity.

3.3
Advantages and Limitations

One main advantage of FBRM is its high statistical robustness, with the measurement counting up to several hundreds of thousands of particles per second, depending on suspension density and particle size. Besides, the measurement principle is not affected by any assumptions about particle shape, like for example in the case of laser diffraction. A morphology change, which can be due to a polymorph transformation, is therefore directly captured in the CLD data. Moreover, FBRM can be used in a wide range of process conditions, both in terms of temperature and pressure and in terms of solid concentration. In principle, there is no upper limit of suspension density that can be measured through FBRM. However, at high suspension densities the measured CLDs do not correlate with particle concentration in a linear way. Finally, FBRM is a count-based technique, which makes the measurement particularly sensitive to fine particles, and it is therefore particularly suited to monitor events like nucleation, breakage, and dissolution which can have a major impact on the final product quality. Consequently, FBRM technology can be considered as a process characterization and optimization tool, suitable for monitoring the particle system dynamics in terms of rate and degree of change of particle number and dimension. This allows the user to understand and quantify the impact of different process parameters on the particulate product.

Besides the aforementioned difficulties in determining the PSD from the measured CLD, other effects may limit the application of the FBRM. The measured CLD can be influenced by the stirring conditions in the crystallizer and the flow field around the probe. Also the size, shape, and number of particles in the suspension can affect the CLD (Worlitschek and Mazzotti, 2003; Kail, Briesen, and Marquardt, 2007; Yu and Erickson, 2008). The most important limitation though is in dealing with transparent particles, where no backscattering or chord splitting may occur (Kail, Briesen, and Marquardt, 2008). In such cases, the optical properties of the solid material play a decisive role and may limit the application of FBRM.

References

Aamir, E., Nagy, Z.K., and Rielly, C.D. (2010) Evaluation of the effect of seed preparation method on the product crystal size distribution for batch cooling crystallization processes. *Cryst. Growth Des.*, **10** (11), 4728–4740.

Abu Bakar, M.R., Nagy, Z.K., Saleemi, A.N., and Rielly, C.D. (2009) The impact of direct nucleation control on crystal size distribution in pharmaceutical crystallization processes. *Cryst. Growth Des.*, **9** (3), 1378–1384.

Barrett, P. and Glennon, B. (2002) Characterizing the metastable zone width and solubility curve using Lasentec FBRM and PVM. *Chem. Eng. Res. Des.*, **80** (A7), 799–805.

Kail, N., Briesen, H., and Marquardt, W. (2007) Advanced geometrical modeling of focused beam reflectance measurements (FBRM). *Part. Part. Syst. Charact.*, **24** (3), 184–192.

Kail, N., Briesen, H., and Marquardt, W. (2008) Analysis of FBRM measurements by means of a 3D optical model. *Powder Technol.*, **185** (3), 211–222.

Kempkes, M., Eggers, J., and Mazzotti, M. (2008) Measurement of particle size and shape by FBRM and *in situ* microscopy. *Chem. Eng. Sci.*, **63** (19), 4656–4675.

Kim, S. *et al.* (2005) Control of the particle properties of a drug substance by crystallization engineering and the effect on drug product formulation. *Org. Process Res. Dev.*, **9**, 894–901.

Lafferrère, L., Hoff, C., and Veesler, S. (2004) *In situ* monitoring of the impact of liquid–liquid phase separation on drug crystallization by seeding. *Cryst. Growth Des.*, **4** (6), 1175–1180.

Li, M.Z. and Wilkinson, D. (2005a) Determination of non-spherical particle size distribution from chord length measurements. Part 1: theoretical analysis. *Chem. Eng. Sci.*, **60** (12), 3251–3265.

Li, M.Z. and Wilkinson, D. (2005b) Determination of non-spherical particle size distribution from chord length measurements. Part 2: experimental validation. *Chem. Eng. Sci.*, **60** (18), 4992–5003.

Liotta, V. and Sabesan, V. (2004) Monitoring and feedback control of supersaturation using ATR-FTIR to produce an active pharmaceutical ingredient of a desired crystal size. *Org. Process Res. Dev.*, **8** (3), 488–494.

Mousaw, P., Saranteas, K., and Prytko, B. (2008) Crystallization improvements of a diastereomeric kinetic resolution through understanding of secondary nucleation. *Org. Process Res. Dev.*, **12**, 243–248.

Ruf, A., Worlitschek, J., and Mazzotti, M. (2000) Modeling and experimental analysis of PSD measurements through FBRM. *Part. Part. Syst. Charact.*, **17** (4), 167–179.

Schoell, J., Bonalumi, D., Vicum, L., Mueller, M., and Mazzotti, M. (2006a) *In situ* monitoring and modeling of the solvent-mediated polymorphic transformation of L-glutamic acid. *Cryst. Growth Des.*, **6** (4), 881–891.

Schoell, J., Vicum, L., Mueller, M., and Mazzotti, M. (2006b) Precipitation of L-glutamic acid: determination of nucleation kinetics. *Chem. Eng. Technol.*, **29** (2), 257–264.

Scott, C. and Black, S. (2005) In-line analysis of impurity effects in crystallization. *Org. Process Res. Dev.*, **9** (6), 890–893.

Togkalidou, T., Braatz, R.D., Johnson, B.K., Davidson, O., and Andrews, A. (2001) Experimental design and inferential modeling in pharmaceutical crystallization. *AIChE J.*, **47** (1), 160–168.

Woo, X.Y., Nagy, Z.K., Tan, R.B.H., and Braatz, R.D. (2009) Adaptive concentration control of cooling and antisolvent crystallization with laser backscattering measurement. *Cryst. Growth Des.*, **9** (1), 182–191.

Worlitschek, J., Hocker, T., and Mazzotti, M. (2005) Restoration of PSD from chord length distribution data using the method of projections onto convex sets. *Part. Part. Syst. Charact.*, **22** (2), 81–98.

Worlitschek, J. and Mazzotti, M. (2003) Choice of the focal point position using Lasentec FBRM. *Part. Part. Syst. Charact.*, **20** (1), 12–17.

Wynn, E.J.W. (2003) Relationship between particle-size and chord-length distributions in focused beam reflectance measurement: stability of direct inversion and weighting. *Powder Technol.*, **133** (1–3), 125–133.

Yu, Z.Q., Chow, P.S., and Tan, R.B.H. (2008) Interpretation of focused beam reflectance measurement (FBRM) data via simulated crystallization. *Org. Process Res. Dev.*, **12** (4), 646–654.

Yu, W. and Erickson, K. (2008) Chord length characterization using focus beam reflectance measurement probe – methodologies and pitfalls. *Powder Technol.*, **185**, 24–30.

4
Turbidimetry for the Estimation of Crystal Average Size

Angelo Chianese, Mariapaola Parisi, and Eugenio Fazio

4.1
Introduction

Turbidimetric techniques have been used since long time for the determination of particle size in suspension. However, this technique is particularly suitable for very small particles from several nanometers to some tens of microns. In fact, many experimental studies (Wallach and Heller, 1964; Maron, Pierce, and Ulevitch, 1963) are focused on latexes, which are submicronic particles, used as reference for crystal size distribution (CSD) measurement techniques. Nevertheless, for continuous monomodal particle distributions and widely separated bimodal distributions, specific turbidity, that is the turbidity per unit volume, may provide a correct location of the weight average size.

In this chapter, firstly a rigorous approach to determine the particle size from the specific turbidity is presented, and then a useful procedure for an estimation of the average size is described with reference to slurries of potassium sulfate crystals.

4.2
Determination of Average Particle Size from Specific Turbidity

Turbidity gives a measure of the attenuation of a light beam crossing a particle suspension:

$$\tau = \frac{1}{l} \ln\left(\frac{I_0}{I}\right) \qquad (4.1)$$

where I_0 and I are the intensities of the incident and transmitted beams, respectively, and l is the length of the optical path. The ratio between the intensities of incident and transmitted light is usually called light absorbance ("ABS") and is here expressed in the logarithmic term, as $\ln(I_0/I)$.

For a suspension of spherical, nonabsorbing, isotropic particles and in the absence of multiple scattering, turbidity can be expressed as a function of the

Industrial Crystallization Process Monitoring and Control, First Edition. Edited by
Angelo Chianese and Herman J. M. Kramer.
© 2012 Wiley-VCH Verlag GmbH & Co. KGaA. Published 2012 by Wiley-VCH Verlag GmbH & Co. KGaA.

number of particles per unit volume, N, and the particle size distribution, as follows:

$$\tau = N \int_0^\infty \frac{\pi D^2}{4} K_{scatt} \cdot f(D) dD \tag{4.2}$$

where $f(D)$ is the normalized number particle size distribution and K_{scatt} is the scattering coefficient, dependent on two parameters: α and m. The parameter α is equal to $\pi(D/\lambda_m)$, which is the ratio of the particle size to the light beam's wavelength, λ_m, whereas m is equal to n_p/n_m, the ratio of the refractive index of the particles to the refractive index of the medium. Summarizing, the general dependence of K_{scatt} from physical parameters is as follows:

$$K_{scatt} = f\left(\frac{D}{\lambda_m}, \frac{n_p}{n_m}\right) \tag{4.3}$$

The determination of the relationship between K_{scatt} and the above-mentioned parameters is a hard task. However, for particles very large compared to the wavelength, as usually occurs in the case of crystal suspensions, the scattering coefficient approaches a constant value independent of m (Kourti, McGregor, and Hamielec, 1991).

Equation (4.2) states that the turbidity of suspension is a function of both the number concentration of particles and the particle size distribution. Since the particle size distribution cannot be estimated by turbidimetric methods, an approximated approach is usually adopted to evaluate the average crystal size of a particle population, under the hypothesis of a monomodal particle size distribution. The method is based on the use of the specific turbidity given by

$$\frac{\tau}{\phi} = \frac{3}{2} \frac{\int_0^\infty D^2 K_{scatt} f(D) dD}{\int_0^\infty D^3 f(D) dD} \tag{4.4}$$

where ϕ is the particle volume fraction. For the monodisperse particle distribution the specific turbidity is simply given by

$$\frac{\tau}{\phi} = \frac{3}{2} \frac{K_{scatt}(D)}{D} \tag{4.5}$$

For practical applications, the volume-surface average diameter, that is the ratio between the third and second moment of the particle size distribution (PSD), usually indicated as D_{32}, can be obtained from specific turbidity measurements of suspension of particles, if the diameter is very large with respect to the adopted wavelength of the light beam ($D \gg \lambda_m$), that is when a constancy of K_{scatt} is attained. A number of researchers used turbidimetric techniques for the determination of the average particle size of polydisperse suspensions with particle diameters in the range from few microns to some tens of microns. Bagchi and Vold (1975) verified that the inverse proportionality between the specific turbidity and the crystal size occurs for large particles. Garcia-Rubio (1987) suggested that Equation (4.5) holds also for nonspherical particles, when D is expressed as the ratio between the third and the second moment of the CSD. However, when the solid concentration of particles, having a specific size distribution, overcomes an upper limit, the multiple scattering becomes important and the linearity between turbidity and solid magma density does not occur any more. This limit represents the maximum solid concentration at which the turbidimetric method may be applied to evaluate the average particle size. To extend the use of this technique to the case of high slurry concentration an automatic dilution may be adopted.

4.3 Procedure to Evaluate Average Crystal Size by Turbidimetry for a High Solid Slurry Concentration

In an industrial crystallizer the solid slurry concentration varies up to 40% by volume and the crystal size usually is in the range between a few microns to several hundreds of microns. In this case the dependence of turbidity on the magma density cannot be linear in the whole range and a dilution of the slurry sample is required to evaluate the average crystal size by turbidimetric measurement. Hereafter, a procedure applied by Cugola *et al.* (2006) to the measurement of the average size of potassium sulfate crystals in aqueous saturated solutions is described. In this work a turbidity cell with an optical path length of around 5 mm was adopted, in order to increase as much as possible the signal of the transmitted light intensity.

The measurement is based on a relationship, to be found, between the specific turbidity and a meaningful average dimension of the crystals, L_{av}. Such a relationship represents the calibration curve of the instrument. In the examined case it was adopted as a characteristic crystal dimension, the second one, which is the reference dimension when the particle size measurement is made, as in this case, by sieving.

First of all, the upper limit of the solid concentration assuring a linearity between turbidity and magma density has to be determined for a slurry sample with the smallest crystal average size. In this particular case a slurry sample with the second dimension of all the crystals smaller than 125 μm was considered to determine the upper limit solid concentration for the measurements. A linear response in terms of ABS vs. M_T was obtained up to a solid concentration of 4%. All the measurements were thus carried out at lower solid concentrations, up to 2%.

In general, the calibration procedure concerns absorbance measurements at various slurry densities for suspensions of different samples of crystals in a close size range. Obviously, the mother liquor of the suspension is at a saturation condition. In particular, the calibration is carried out in two steps:

1) For each examined sample of crystals, with a specific average size L_{av}, a series of measurements are made by changing the crystal suspension density and a couple of data (ABS, M_T) is obtained at each solid concentration. The ratio ABS/M_T is proportional to the specific turbidity, according to Equation (4.1),

and, by assuming on the first approximation a monodisperse distribution, it is a unique function of the average crystal size (see Equation (4.5)). Accordingly, the linear fitting with zero intercept between ABS and M_T gives a proportionality coefficient, say K_p. This coefficient depends on both L_{av} and K_{scatt}, which is also a function of the crystal size, and as a consequence it is affected by the relationship between K_{scatt} and L_{av}, expressed by Equation (4.3). Just one data (ABS, M_T) achieved by an experimental measurement could be enough to determine the K_p value corresponding to a specific average diameter. However, it is advisable, for the sake of accuracy, to derive this latter value at least from three different data obtained by runs performed at three different solid mass concentrations, that is, by dilution of the original sample.

2) The average crystal size of each sample adopted for calibration purpose is plotted against the proportionality coefficient, obtained as reported above, and the linear interpolation equation is assumed as the calibration curve of the measurement technique.

In Figures 4.1 and 4.2, the results of the first and the second step of the suggested procedure, respectively, are shown.

Figure 4.1 shows the linear relationship obtained for all the examined slurry samples in the examined solid concentration range. From the best linear fitting of the data the values of K_p are obtained, as slopes. These values are then reported, in abscissa, vs. the mass average crystal size of each sample, determined by sieving, in Figure 4.2: the linear relationship can be adopted as the calibration curve for the measurement procedure.

For both the figures the data can be reasonably interpolated by linear fittings. In particular, the linear fitting of the data in Figure 4.2 gives the following equation:

$$L_{av} = 275.0 - 11.35 K_p \tag{4.6}$$

where L_{av} is in micron, with the correlation index equal to 0.98.

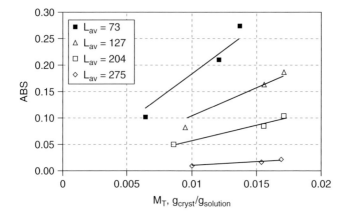

Figure 4.1 ABS vs. M_T for different L_{av} (μm). The slope of each straight line is the proportionality coefficient, K_p.

4.3 Procedure to Evaluate Average Crystal Size by Turbidimetry for a High Solid Slurry Concentration

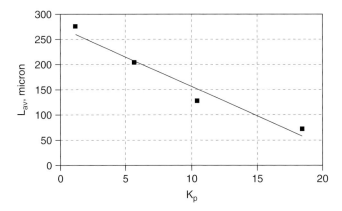

Figure 4.2 Calibration curve for the examined potassium sulfate system.

It is interesting to observe that, due the changeable value of K_{scatt}, K_p varies much more than L_{av}; in fact the range of the average size of the examined slurries is 73–275 μm, whereas the proportionality coefficients changes between 19.2 and 1.2, respectively.

Once the calibration curve is available, it is possible to estimate the average size of each sample by a two-step procedure: (i) to carry out some measurements of the light absorption, ABS, at a different slurry density, M_T, of the examined crystal sample; (ii) to determine the proportionality coefficient, K_p, as the slope of the linear correlation between ABS and M_T; and (iii) to estimate L_{av} from the determined value of K_p by means of the calibration curve.

To validate the proposed procedure, many checks on the crystal samples having known CSD were made. A typical example is shown hereafter. A K_2SO_4 crystal sample, having an average diameter of 197.4 μm, was put in aqueous saturated solutions at three different slurry density values. The absorption values derived by the turbidity measurements are plotted in Figure 4.3.

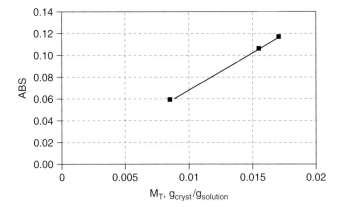

Figure 4.3 Linearity between ABS and M_T for the new sample.

A very good linearity was obtained (correlation index > 0.99) with a slope equal to 7.06. Putting in Equation (4.6) a proportionality coefficient value of 7.06 an average crystal size equal to 194.8 μm is calculated; thus, in this particular case the turbidimetric measurement method gave an accuracy of 1.3%.

4.4
Conclusion

Turbidimetry may be a useful technique to measure the average size of a crystal sample. The measurement technique suggested here is more suitable for plant laboratory either for the not high accuracy or because of the need of a careful calibration procedure. In this kind of application, the use of this technique can lead to quick measurements of the average size of the crystal population in a slurry. In fact, in it, it is not required to dry the samples in order to apply the sieving technique, but its prerequisite is just the dilution of the withdrawn slurry sample with a saturated solution in order to merge the size range where the absorbance is a linear relationship of the suspension mass density.

References

Bagchi, P. and Vold, R.D. (1975) A simple method for determination of the average particle size of coarse suspensions from measurements of apparent specific turbidity. *J. Colloid Interface Sci.*, **53**, 194–201.

Cugola, M., Parisi, M., Chianese, A., and Fazio, E. (2006) Prediction of average crystal size by turbidimetry. CGOM 7, August 2006, Rouen.

Garcia-Rubio, L.H. (1987) *Particle Size Distribution–Assessment and Characterization*, ACS Symposium Series, Vol. 332, ACS, Washington, DC, pp. 161–178.

Kourti, T., McGregor, J.F., and Hamielec, A.E. (1991) *Particle Size Distribution II*, ACS Symposium Series, Vol. 472, ACS, Washington, DC, pp. 2–19.

Maron, S.H., Pierce, P.E., and Ulevitch, I.N. (1963) Determination of latex particle size by light scattering: IV. Transmission measurements. *J. Colloid Sci.*, **18**, 470–482.

Wallach, M.L. and Heller, W. (1964) Experimental investigations on the light scattering of colloidal spheres: VI. Determination of size distribution curves by means of turbidity spectra. *J. Phys. Chem.*, **68**, 924–930.

Further Reading

Chianese, A., Bravi, M., Cugola, M., and Mascioletti, A. (2005) Turbidimetri di processo per il controllo di qualità nella cristallizzazione industriale. Proceedings of the IV Congress on Metrology and Quality, February 2005, Torino, pp. 93–96.

Chianese, A., Bravi, M., and Mascioletti, A. (2005) Metodo e strumento optoelettronico per la misura della solubilità di sostanze chimiche, Italian Patent no. RM2005A000182.

Elicabe, G.E. and Garcia-Rubio, L.H. (1988) The selection of the regularization parameter in inverse problems: estimation of particle size distribution from turbidimetry. *Polym. Mater. Sci. Eng.*, **59**, 165–168.

5
Imaging
Herman J.M. Kramer and Somnath S. Kadam

5.1
Introduction

The crystal size distribution (CSD) of the product of a crystallization or a precipitation process is the main property, determining the product quality and the product performance. It also affects the efficiency of the downstream solid liquid separation steps like filtration and drying. As already discussed in this book, a number of CSD measurement devices are available such as laser diffraction (LD), ultrasonic attenuation (UA), and focused beam reflectance measurement (FBRM). All of these measurement devices suffer from major difficulties during *in situ* monitoring of the CSD and do not provide reliable size information especially when the shape of the particles deviates from the ideal, spherical shape. In addition, for both the LD and the UA instruments a reconstruction of the size distribution from the measured signals is required which is complicated for nonideal (multimodal) distributions too.

Video microscopy has been shown to be a valuable alternative for these CSD measurement tools as it does not suffer from these problems. The direct observation of the particles makes the interpretation of the data intuitive. It is important to note that the size and shape of the crystals are obtained without additional assumptions of crystals' shape or of their size distribution. In addition, the 2D crystal shape information obtained allows the characterization of both the size and the shape of the crystals in a single measurement, without extensive calibration procedures, simplifying the experiments and reducing the cost of the instrumentation. The shape information from the image analysis could be vital to monitor and control the type of polymorph or the polymorph transformation for chemical systems showing different crystal structures (Larson, 2007).

To derive quantitative information from the images requires image segmentation. This means that the objects of interest need to be separated from the background. For this purpose extensive techniques are needed to improve the image quality, perform background correction, improve the sharpness of the images correct for overlapping and boundary problems, and so on. The image analysis is already well established and a number of commercial instruments are available both for off-line and for *in situ* analysis.

Most of the commercial instruments solve the segmentation problem by imaging the particulate slurry as it passes through a specially designed flow cell under controlled hydrodynamic and lighting conditions (Allen, 2003). Examples of these off-line particle size and shape analysis instruments are Malvern's Sysmec FPIA 3000, Beckman Coulter's RapidVUE, and Sympatec's QICPIC. Although these instruments are capable of delivering a size and shape distribution, they are not considered here, since they require sampling and sample pretreatment. Taking representative samples from particle suspensions is notoriously difficult (Allen, 1996, 2003), time consuming, and problematic (not robust). There are also a number of process sensors that allow for image analysis directly in the process, like the Mettler Toledo particle vision and measurement (PVM) probe (*www.mt.com*), the *in situ* particle viewer (ISPV) of Perdix Analytical Systems (*www.perdix.nl*), the process image analyzer (PIA) of MessTechnik Schwartz, and the EZProbe D25 developed at the University of Lyon 1 (Gagniére, 2009). The application of *in situ* imaging for monitoring and control of crystallization processes however still suffers from problems with the segmentation. Due to varying background intensities, overlapping particles, and sharpness of the images, quantitative information from these sensors is still not very reliable and needs more development.

In the next section, an overview of the available literature about the application of *in situ* image analysis systems is given. In addition, a comparison with other CSD measurement instruments and a discussion on specific aspects of the imaging sensor design are reported. Finally, a number of case studies are discussed, which show the opportunities and the limitations of the application of imaging for the *in situ* monitoring of industrial crystallization processes.

5.2
Literature Overview

One of the important attributes which defines the quality of the crystalline product is the CSD. Hence most of the efforts in crystallization control are directed toward obtaining products with predefined CSD (Fujiwara, Chow, and Braatz, 2002). In this respect, an important aspect is the definition of the crystal size, as the shape of the crystals can vary remarkably for different compounds or even for the same compound under different process conditions. This means that during the monitoring of the CSD, the differences in the shapes should be taken into account. This presents a big challenge for the CSD measurement instruments.

The commonly used size measurement instruments are based on LD, laser backscattering, and ultrasound extinction. LD records the diffraction pattern created by the crystals and tries to determine the distribution of diameters of spherical particles which would have similar diffraction patterns (see Chapter 2). In a LD measurement, the shape of the original crystals is not taken into account. Due to the spherical models used to interpret the data and the orientation of crystals, LD tends to overestimate the broadness of the diameter distribution of high aspect ratio particles (Naito et al., 1998). FBRM measures the chord length distribution (CLD) of the crystals, which can in principle be inverted to obtain the CSD but the inversion requires assumptions about the particle shape and is very ill-conditioned (Worlitschek, Hocker, and Mazzotti, 2005). The use of ultrasound extinction to measure the CSD requires a lot of additional information which might not be readily available or relies on calibration against another instrument (in most cases LD). In both cases however the shape of the particles cannot be taken into account. Although most of the instruments work well for nearly spherical particles, they suffer when particles are not spherical, especially when they have a high aspect ratio.

Most of these issues with measurement of the CSD can be effectively tackled by using imaging. One of the strongest advantages of image analysis is that it is a direct observation technique. Hence no assumptions have to be made about the particle shape and also no inversion techniques are needed to obtain the particle size distribution[1] CSD (Patience and Rawlings, 2001). The other advantage is that image analysis can be used to measure the 2D CSD. Furthermore, direct observation during imaging makes it possible to distinguish between different crystallization phenomena like dissolution and agglomeration both of which lead to a decrease in particle number (Patience, Dell'Orco, and Rawlings, 2004). Although imaging has many advantages, one of the big limitations is that it is sensitive to quality of the images. It performs better when images of the crystals are taken off-line and the crystals are arranged on a single plane without any overlaps. Images taken *in situ* present additional challenges, due to varying background intensity, which makes image segmentation more difficult. Also difficulties arise due to limited depth of field, a parameter associated with the instrumentation. A combination of limited depth of field and the random orientation of the crystals during *in situ* imaging leads to variations in edge thickness (blurred edges). Even if the crystals are oriented perpendicular to the lens, crystals in a single snapshot may have different clarities due to variations in the distance with respect to the lens.

In spite of its limitation, the advantage of direct observation that image analysis offers makes it potentially one of the best tools for online crystallization monitoring and control.

Simon, Nagy, and Hungerbuhler (2009) used endoscopy-based bulk video imaging (BVI) to monitor nucleation onset and for meta-stable zone width determination of potash alum hydrate in water. They captured images from an external endoscope, converted them to gray format, and determined its mean gray intensity (MGI). An increase in the MGI of images above a certain threshold value indicates the onset of nucleation. Although a lot of qualitative information is obtained through BVI, it is not suitable for determining the CSD as the resolution of image is low compared to crystal size, optical system of endoscope introduces image distortion,

---

1) Strictly speaking this is only true for 3D imaging as in 2D imaging still assumptions are needed to convert the measured projected area into the shape in three dimensions.

and the moving objects appear as traces. Calderon de Anda *et al.* (2005c) described multiscale image segmentation for analysis of images obtained online during batch cooling crystallization of L-glutamic acid. They used a prototype imaging system developed by GlaxoSmithKline (GSK). Although image segmentations is one of the steps to obtain quantitative information of the CSD, the authors limit themselves to demonstrating the segmentation methodology only. They demonstrate only the multiscale image segmentation for images captured at very low crystal concentrations, resulting in images without overlapping crystals and well-defined edges. Patience and Rawlings (2001) used similar segmentation strategy in combination with a stop-flow through cell placed under a microscope to monitor transient crystal shape changes of sodium chlorate during suspension crystallization in the presence of a habit modifier. They monitored the box area evolution in time to understand the change in crystal habit during the process. *Box area* is defined as the ratio of the area of an object to the area of the minimum size bounding box of the object. They allowed preferred orientation of the crystals in the flow through cell by stopping the flow for a few seconds. This enabled better image quality and also avoided images of crystals in random orientation. In spite of the stop-flow operation, they had to set the light thresholding and focus manually for each image in order to obtain better segmentation of the image.

One of the first quantitative applications of imaging in determining the particle size distribution comes from the field of coal liquefaction. Kaufman and Scott (1994) introduced and validated a fluorescence imaging method which enables *in situ* visualization and sizing of coal particles in a liquid fluidized bed. The liquid phase was made fluorescent by the addition of a fluorescent dye while the coal particles were opaque. This contrast enabled edge detection quite easily with the help of the gray-level thresholding method at low solid concentration. At high solid concentrations, however, manual intervention was required to determine which of the image was good for sizing due to high degree of overlapping particles. In crystallization, however, one of the earliest quantitative image analyses was done off-line by Puel, Marchal, and Klein (1997). Amongst the two new "tools" they developed, one was experimental which could measure characteristic sizes of the examined crystal by image analysis. They used it to analyze habit transient phenomena during continuous crystallization of hydroquinone with an additive. They limited themselves to 2D image analysis by assuming the breadth of the rod-like crystals to be equal to its width. Moreover, the image analysis was done off-line after collecting the sample. They observed that the image analysis was challenging as the crystals had the tendency to overlap and therefore a good dispersion method was necessary to disperse the crystals on the stage below the microscope. Larsen, Rawlings, and Ferrier (2007) went a step further and presented a model-based recognition algorithm which was designed to extract size and shape information from *in situ* crystallization images. They compared the projected areas of the crystal determined by the algorithm with the projected areas determined by a human operator to establish the effectiveness of the algorithm. They observed that the algorithm starts suffering as the particle concentration goes on increasing. Like most of the image analysis studies, the algorithm was not very effective with crystals having low contrast edges and when the crystals were overlapping. But nevertheless, the work was one of the first steps toward analyzing images obtained *in situ* during crystallization processes and the algorithm is fairly accurate at low particle concentrations.

An alternative approach for obtaining the 2D CSD is presented by Kempkes, Eggers, and Mazotti (2008) and Eggers, Kemkes, and Mazzotti (2008). Kempkes *et al.* presented an algorithm which gives the 2D axis length distribution (ALD) based on images collected *in situ* by a video microscope. To determine the ALD, the crystals are distinguished from the background by using the method presented by Calderon de Anda *et al.* (2005a, 2005b and 2005c). Then an ellipsoid with the same area of the examined crystal is fitted to the crystal and its major and minor axis, *a* and *b*, are determined. Eggers *et al.* present a method for restoring the 2D CSD based on the collected ALD by using a genetic algorithm. They demonstrate this approach by simulation as well as experimentally for a very dilute solution of carbon fiber particles. Kempkes, Vetter, and Mazotti (2010) presented additional work in which they took a step further by presenting a method to obtain the 3D particle size distribution from suspension. They developed a setup which uses a single camera coupled with two mirrors to take images of the same crystal simultaneously from two orthogonal directions. From these images, they determined the 4D ALD and then finally obtained the 3D CSD with the help of the genetic algorithm developed by Eggers, Kemkes, and Mazzotti (2008). They applied this method during crystallization of ascorbic acid from methanol and L-glutamic acid from water and claimed that the determined 3D CSD agrees reasonably with the one measured with the help of an electron microscope. Although the images used in this work were taken at low particle concentration in a flow through cell, the work opens new directions for obtaining 3D CSDs.

5.3
The Sensor Design

Despite the fast developments sketched above, robust and efficient segmentation is still a challenge for *in situ* measurement systems. Parts of the problem are related to the instrument, such as type of illumination, optical design, resolution of the image sensor, and so on. In addition, a number of other actors play a role in the retrieval of a robust statistically significant CSD from *in situ* imaging. Figure 5.1 shows some typical layouts for imaging systems, suitable for *in situ* monitoring of the size and shape of a particle suspension. The system contains the following main components.

- Light source with (fiber) optics to illuminate the particles in the measurement volume. In general a flash light with short illumination times is used to obtain bright images. Short illumination times help in avoiding image blurring which is caused by movement of the particles under continuous light illumination and also helps in avoiding heating up of the process liquid.

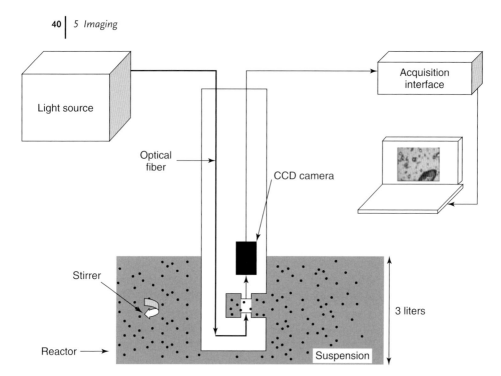

Figure 5.1 General layout of an *in situ* imaging system.

- The detection system consisting of the camera and the optical components used to project the image on the camera, mostly a CCD or a high speed CMOS camera.
- Image analysis system, where the images are converted into distributions of size and shape using a segmentation procedure. As mentioned before, the quality of the images in terms of sharpness, uniformity of the background, overlap of the particle images, and so on determines to a large extent which approach will be followed.

The quality of the images is mainly determined by the type of illumination and by the sensitivity and resolution of the used camera. Fiber optics used to be able to locate the detector outside the reactor might affect the quality of the images. The design of the instruments will not be discussed in detail here. Only a few points concerning the type of illumination, the statistics of the CSD, and the image analysis will be discussed in the next sections.

5.3.1
Optics and Illumination

One of the major factors which determine the geometry of the instrument and the quality of the images is the type of illumination. Front-end illumination illuminates the particles on the same side as they are detected by measuring the reflected light from the particles (see Figure 5.2a). One of the main advantages of this configuration is the simplicity of the probe design, in which the light illuminating the slurry

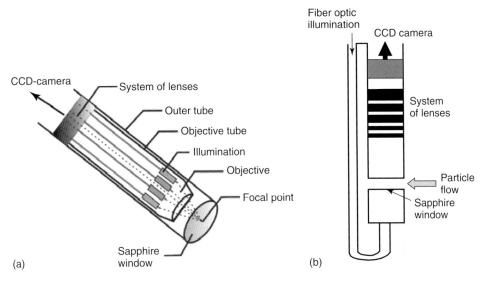

Figure 5.2 (a) Layout of a front-end illumination probe (PVM sensor manufactured by Mettler Toledo/Lasentec, Redmond, WA, USA) and (b) layout of a transmission probe system (PIA system manufactured by Messtechnik Schwartz, partner of Sequip – Sensor and Equipment GmbH, Dusseldorf, Germany).

and the reflected light from the crystals pass through the same window. Also this configuration allows in principle for higher particle concentration to be analyzed, but at higher particle concentration the particles further away from the probe are shielded by the particles closer at the probe window. Thus, the drawback of this is that the sample volume becomes dependent on the particle concentration. Another important drawback of this configuration is the low quality of the images, which is caused by the fact that measured particles have nonuniform colors and intensities, and have often poorly defined outlines, which make threshold methods ineffective and cause problems in algorithms to close the particle outlines. In addition, as the illuminating light beam is not nicely parallel, but mostly originates from a ring-shaped light source, the size of the particles will depend on the distance and position of the particle with respect to the sensor window.

For the illumination in transmission a more complicated probe design is needed (see Figure 5.2b), but this allows us to obtain a good image quality as is illustrated in Figure 5.3.

All off-line instruments and also a number of *in situ* sensors use this type of illumination. In the probe the illumination is separated from the camera allowing a well-defined parallel measurement beam avoiding disturbing effects of the position on the optical size of the particles. Besides the more complicated design, the major drawbacks of this design are the hydrodynamic conditions of the particles in the slit and the low dynamic range of particle concentration that can be handled. The

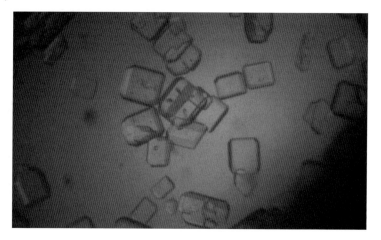

Figure 5.3 *In situ* images of an ammonium sulfate crystals during a cooling crystallization.

design of the probe and especially the width of the slit should allow a representative flow through the slid of the suspension, containing an unclassified particle flow with the same concentration as in the process. This means that the position and orientation of the probe with respect to the main flow pattern in the process becomes important. To allow a fast refreshment of the suspension in the slit a positioning close to the stirrer or an inlet point is needed. Classification in the slit should be avoided in any case and thus the flow rate and the flow direction in the slit should be comparable to the flow rate around the probe. This condition constrains the width of the slit. At low slit widths, the refreshment rate of the particles will be slowed down and can become size dependent. In those cases the measured particles are not representative anymore for the particles in the process. At large slit gaps, that is, a large optical path length, overlapping and shielding might occur at higher particle concentrations. In addition, particles present in front of or behind the focal plane will show up as unsharp images or as disturbances in the image. So a compromise should be found between a wide slit for low concentrations and a narrow one for large particle concentrations with the risk of classification and low refreshment rates. These constrains limit the particle concentration to about a few volume percent, depending on the resolution.

5.3.2
The Camera System and the Resolution

Nowadays high resolution CCD of CMOS cameras are available for imaging systems. Theoretically the spatial resolution of the particles is limited by the wavelength of the adopted light. The use of blue light gives only a small increase in this resolution. Most probes however do not work at this theoretical minimum resolution because of some practical limitations. High resolution means high magnification which has a number of disadvantages. First of all, the depth of field decreases at higher magnifications, reducing the measurement volume. This means that the larger particles are difficult to be captured, while most of the small particles will be present outside the measurement volume, that is, the region where sharp images can be obtained. Postprocessing of the images can improve the quality to a certain degree (Simon, Nagy, and Hungerbuhler, 2009), but mostly requires computer intensive treatment of the data, which is not very suitable for the online monitoring and control. Secondly, the capture of the small particles at high resolution requires very short illumination times. To obtain sharp images of particles that pass the measurement volume with a maximum linear velocity of 5 m/s with a resolution of 1 μm requires a shutter or illumination time of less than 200 ns. Although in some off-lines instruments this short illumination times can be realized (for instance in Sympatec's QICPIC), it is generally not the case for *in situ* probes.

5.3.3
Image Analysis

Extraction of meaningful information from an image involves several steps like image preprocessing to remove noise, segmentation, and characterization of the identified object. Image segmentation is the process of assigning a label to every pixel in an image, such that pixels with the same label represent the same object or its parts (da Costa and Cesar, 2001; Barbieri *et al.*, 2011). It is the most challenging part of image analysis and hence we will focus on image segmentation in this section. Among the several segmentation methods (Pal and Pal, 1993), two of the methods commonly used during crystallization to separate crystals from its background are described hereafter.

1) *Edge detection*: Segmentation by edge detection can be achieved by locating points of abrupt changes in pixel intensity values of a digital image. Among the several available edge detection methods, Canny edge detection works in most scenarios (Maini and Aggarwal, 2009). The performance of the technique is dependent on three parameters, viz. σ, which is the standard deviation for the Gaussian filter used to convolve the image, and the threshold pixel intensity values T1 and T2 (Canny, 1986). Smaller σ causes less blurring and allows detection of small, sharp lines while larger σ causes more blurring smearing out values of pixel over a large image area. The use of two threshold with hysteresis allows more flexibility than in a single threshold approach (Canny, 1986).
Canny edge detection was used by Calderon de Anda, Wang, and Roberts (2005a) in their multiscale image segmentation strategy for analyzing images of L-glutamic acid obtained during batch crystallization. They performed edge detection at two scales. On the first scale they tried to capture image features which certainly were from particle edges, and on the second scale they tried delineation of edges while accepting some noise. The edges detected at these two scales were combined to do image segmentation.
Figure 5.4 shows the results obtained by applying the multiscale image segmentation method on images of L-glutamic acid crystals obtained *in situ*

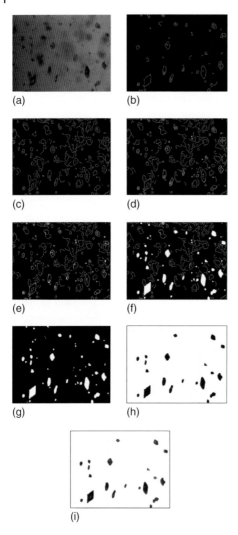

Figure 5.4 The segmentation method applied on a sample *in situ* image with highly irregular pixel intensity (Calderon De Anda, Wang, and Roberts, 2005a). (a) Original image, (b) edges detected at the first scale, (c) edges detected at the second scale, (d) edges of the first and second scales, (e) morphological closing on image (d), (f) region-filling on image (e), (g) morphological opening on image (f), (h) segmented particles after removing those with less than 200 pixels from image (g), and (i) segmented particles with the original gray-scale intensity superimposed.

during batch crystallization. The figure indicates that the method is successfully applied in segmenting the crystals in the image. However, it must be noticed that in this case the crystal concentration in the image is very low and there are no overlapping crystals. In case many instances of overlapping exist, correction techniques need to be applied (Larson, 2007).

2) *Model-based object recognition*: Model-based image recognition is based on detecting groupings and structures in the images that are likely to be invariant over wide viewpoints. Examples of grouping and structures are connectivity, collinearity, parallelism, texture properties, and so on (Lowe, 1987). Once detected, these groupings can be matched with corresponding structures of one of the predefined models to enable segmentation. Larsen, Rawlings, and Ferrier (2007) applied the model-based object recognition for segmentation of crystals in the images collected during cooling crystallization of α-glycine. They claimed that the algorithm can be applied to images of crystals of any shape, provided the shape can be represented as a wire-frame model. The effectiveness of the algorithm was demonstrated by comparing the algorithm results with those obtained by manual human analysis of 300 *in situ* images acquired at different solid concentrations during α-glycine cooling crystallization (see Figure 5.5).

3) It was observed that at low solid concentrations, the algorithm could recognize half of the crystals identified by a human while at medium to high concentrations, the algorithm could identity one third of the crystals identified by a human operator. On the other hand, the number of falsely recognized crystals

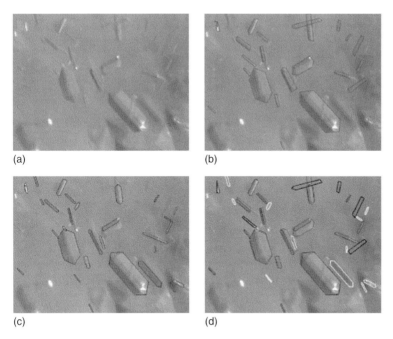

Figure 5.5 Illustration of the comparison between human operator results and M-SHARC results. (a) Original image. (b) Outlines of crystals identified by human operator. (c) Outlines of crystals identified by M-SHARC. (d) Result of comparison between human outlines and M-SHARC outlines. Crystals identified as false positives are outlined in white while those identified as misses are outlined in black.

increased from one fifth at low concentration to almost half at high particle concentration.

5.3.4
Statistics

Imaging is a number-based measurement technique. In order to get a statistical relevant CSD a large number of particles need to be analyzed (Allen, 1996). As a rule of thumb depending on the width of the distribution and the required accuracy, 400–10 000 particles need to be counted per size interval, which means 12 000–300 000 particles per CSD. This would require a measurement time of 40–1000 s assuming 10 particles per image and a frame rate of 30 images per second. For a two-dimensional CSD in which the number of particles is plotted as a function of both the size and shape of the particles, many more particles are needed to acquire the same accuracy. It might be clear that a huge amount of data has to be treated, challenging both the data transfer rates and the data storage capacity of the computer system. It is therefore important to try to reduce the amount of data to be transferred and stored. For high quality images with a more or less constant background, this can be achieved automatically by background subtraction and thresholding and converting the data into black and white images. However, at lower image quality, especially with varying background thresholds or when postprocessing has to be done to achieve acceptable image quality, detailed data have to be stored.

5.4
Application of In Situ Imaging for Monitoring Crystallization Processes

5.4.1
Example 1

One of the earliest quantitative work in image analysis during crystallization was done by Puel, Marchal, and Klein (1997), where characteristic crystal sizes were measured off-line to analyze habit transient phenomena during continuous crystallization of hydroquinone in the presence of an additive. Hydroquinone crystals exhibit a rod-like habit which can be described with length, breadth, and width, which is assumed equal to its breadth. The crystal samples were collected and analyzed off-line on a stage under a microscope connected to a camera. Using a constant shape factor the number distribution obtained by image analysis was converted to the mass distribution and compared with the mass distribution obtained by sieving as shown in Figure 5.6.

As can be seen from Figure 5.6, both the distributions have comparable widths with the coefficient of variation slightly lower for image analysis. To obtain the above result, the authors counted a minimum of 1000 particles between 10 µm and 1 mm, which was twice the number recommended in the literature (Allen, 1996).

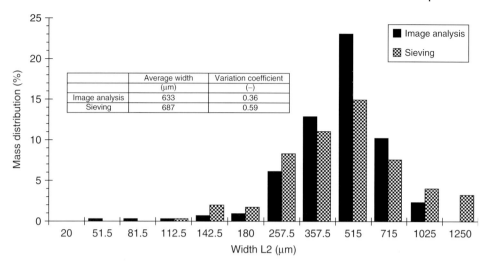

Figure 5.6 Comparison of two mass distributions obtained by sieving and image analysis for hydroquinone crystals.

5.4.2
Example 2

Wang, Calderon De Anda, and Roberts (2007) used online imaging and image analysis for real-time measurement of the length and width of the needle-shaped β L-glutamic acid crystals. Simultaneous estimation of growth rate of different faces could be valuable to monitor the evolution of the crystal shape. During their experiments, they heated the suspension of the crystals above the saturation temperature but allowed a few particles to survive the dissolution. The survived particles acted as seeds during the cooling phase. The process was monitored by an imaging system developed by GSK (Calderon De Anda, Wang, and Roberts, 2005a; Calderon De Anda *et al.*, 2005b) with a pixel resolution of 640 × 480. The camera was situated outside the reactor wall and images were acquired at the frequency of 30 images/s.

The time evolution of the length and width of the crystals is plotted in Figure 5.7.

Each point in the figure is averaged over 300 images collected and analyzed for each point. Due to the physical limits imposed by the lenses on the magnification of the image the detection limit of the image analysis system was 30 µm. The estimated growth rates based on the sizes obtained by image analysis was compared with the growth rates values based on single crystal experiments reported in the literature (Kitamura and Ishizu, 2000). The value reported by Kitamura *et al.* in the literature was found to be lower than that reported by Wang, Calderon De Anda, and Roberts (2007) who attributed the deviation to different hydrodynamic conditions in the two setups.

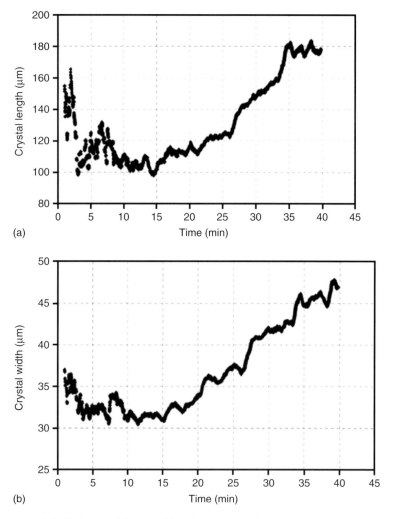

Figure 5.7 Evolution of the crystal length (a) and width (b) for the needle-shaped glutamic acid crystals. Each point represents the mean values in the previous 60 s (i.e., 300 images). The time $t = 0$ min corresponds to the onset temperature, that is, 70.7 °C.

5.5
Conclusions

Image analysis has become a powerful technique to monitor the development of the product quality in crystallization processes. Due to the improvements of the optics, the illumination system and the high speed and high resolution camera systems that are available nowadays, the quality of the information has tremendously improved. Most importantly, this valuable information on the development of the size and the shape of the crystals becomes available in "real time" during the crystallization process itself. In addition, apart from shape and size information, image analysis also gives information on different phenomena like dissolution/agglomeration which other CSD measurement techniques might not recognize. *In situ* imaging has therefore evolved as a powerful tool for operators to optimize the crystallization process. The application of this sensor for the model-based control of crystallization processes is in principle also possible but requires a robust estimation of the CSD and the shape distribution in real time. In practical situations this quantitative analysis has not been yet achieved due to limitations of the image quality, the low dynamic range of particle concentration that can be handled by the sensor, and a lack of robustness of the image analysis algorithms. More developments are needed in these areas.

References

Allen, T. (1996) *Particle Size Measurement*, 5th edn, Springer, Berlin.

Allen, T. (2003) *Powder Sampling and Particle Size Determination*, Elsevier, Amsterdam.

Barbieri, A.L. *et al.* (2011) An entropy based approach to automatic image segmentation of satellite images, *Phys. A: Stat. Mech. Appl.*, **390** (3), 512–518.

Calderon De Anda, J., Wang, X.Z., and Roberts, K.J. (2005a) Multi-scale segmentation image analysis for the in-process monitoring of particle shape with batch crystallizers. *Chem. Eng. Sci.*, **60** (4), 1053–1065.

Calderon De Anda, J., Wang, X., Lai, X., and Roberts, K. (2005b) Classifying organic crystals via in-process image analysis and the use of monitoring charts to follow polymorphic and morphological changes. *J. Process Control*, **15** (7), 785–797.

Calderon De Anda, J., Wang, X.Z., Lai, X., Roberts, K.J., Jennings, K.H., Wilkinson, M.J., Watson, D., and Roberts, D. (2005c) Real-time product morphology monitoring in crystallization using imaging technique. *AIChE J.*, **51**(5), 1406–1414.

Canny, J. (1986) A computational approach to edge detection. *IEEE Trans. Pattern Anal. Mach. Intell.*, **8**(6), 679–698.

da Costa, L.F. and Cesar, R. (2001) *Shape Analysis and Classification: Theory and Practice*, CRC Press, Boca Raton, FL.

Eggers, J., Kemkes, M., and Mazzotti, M. (2008) Measurement of size and shape distributions of particles through image analysis. *Chem. Eng. Sci.*, **63** (22), 5513–5521.

Fujiwara, M., Chow, P.S., and Braatz, R.D. (2002) Paracetamol crystallization using laser backscatter ring and ATR-FTIR spectroscopy, metastability, agglomeration and control. *Cryst. Growth Des.*, **2**, 363–370.

Gagniére, E., Mangin, E., Puel, R., Rivoire, A., Monnier, O., Garcia, E., and Klein, J.P. (2009) Formation of co-crystals: Kinetic and thermodynamic aspects. *J. Cryst. Growth*, **311**(9), 2689–2695.

Kaufman, E.N. and Scott, T.C. (1994) Insitu visualization of coal particle distribution in a liquid fluidized bed using fluorescence microscopy, *Powder Technol.*, **78** (3), 239–246.

Kempkes, M., Eggers, J., and Mazotti, M. (2008) Measurement of particle size and shape by FBRM and *in situ* microscopy. *Chem. Eng. Sci.*, **63** (19), 4656–4675.

Kempkes, M., Vetter, T., and Mazotti, M. (2010) Measurement of 3D particle size distributions by stereoscopic imaging. *Chem. Eng. Sci.*, **65** (4), 1362–1373.

Kitamura, M. and Ishizu, T. (2000) Growth kinetics and morphological change of polymorphs of L-glutamic acid. *J. Cryst. Growth*, **209** (1), 138–145.

Larson, P. (2007) Computer vision and statistical estimation tools for *in situ*, imaging-based monitoring of particulate populations. PhD thesis. University of Wisconsin, Madison.

Larsen, P.A., Rawlings, J.B., and Ferrier, N.J. (2007) Model based object recognition to measure crystal size and shape distributions from *in situ* video images. *Chem. Eng. Sci.*, **62** (5), 1430–1441.

Lowe, D. (1987) Three dimensional object recognition from single two-dimensional images. *Artif. Intell.*, **31**, 355–395.

Maini, R. and Aggarwal, H. (2009) Study and comparison of various image edge detection techniques. *Int. J. Image Process.*, **3** (1), 1–11.

Naito, M., Hayakawa, O., Nakahira, K., Mori, H., and Tsubaki, J. (1998) Effect of particle shape on the particle size distribution measured with commercial equipment. *Powder Technol.*, **100**, 52–60.

Pal, R. and Pal, S.K. (1993) A review on image segmentation techniques. *Pattern Recognit.*, **26**, 1277–1294.

Patience, D.B., Dell'Orco, P.C., and Rawlings, J.B. (2004) Optimal operation of a seeded pharmaceutical crystallization with growth-dependent dispersion. *Org. Process Res. Dev.*, **8** (4), 609–615.

Patience, D.B. and Rawlings, J.B. (2001) Particle shape monitoring and control in crystallization processes. *AIChE J.*, **47** (9), 2125–2130.

Puel, F., Marchal, P., and Klein, P. (1997) Habit transient analysis in industrial crystallization using two dimensional crystal sizing technique. *Chem. Eng. Res. Des.*, **75** (2), 193–205.

Simon, L., Nagy, Z.K., and Hungerbuhler, K. (2009) Endoscopy-based *in situ* bulk video imaging of batch crystallization processes. *Org. Process Res. Dev.*, **13** (6), 1254–1261.

Wang, X.Z., Calderon De Anda, J., and Roberts, K.J. (2007) Real-time measurement of the growth rates of individual crystal facets using imaging and image analysis: a feasibility study on needle-shaped crystals of L-glutamic acid. *Chem. Eng. Res. Des.*, **85** (7), 921–927.

Worlitschek, J., Hocker, T., and Mazzotti, M. (2005) Restoration of PSD from chord length distribution data using the method of projections onto convex sets. *Part. Syst. Charact.*, **22** (2), 81–98.

Further Reading

Monnier, O., Fevotte, G., Hoff, C., and Klein, J.P. (1997) Model identification of batch cooling crystallizations through calorimetry and image analysis. *Chem. Eng. Sci.*, **52** (7), 1125–1139.

Neumann, A.M. and Kramer, H.J.M. (2002) A comparative study of various size distribution measurement systems. *Part. Syst. Charact.*, **19**, 17–27.

Presles, B., Debayle, J., Fevotte, G., and Pinoli, J.-C. (1997) Novel image analysis method for *in situ* monitoring the particle size distribution of batch crystallization processes. *J. Electron. Imaging*, **19**, 031207-1–031207-7.

Scholl, J., Bonalumi, D., Vicum, L., Mazzotti, M., and Muller, M. (2006) *In situ* monitoring and modeling of the solvent-mediated polymorphic transformation of L-glutamic acid. *Cryst. Growth Des.*, **6** (4), 881–891.

6
Turbidimetry and Nephelometry

Angelo Chianese, Marco Bravi, and Eugenio Fazio

6.1
Introduction

Turbidimetry is a very simple technique to evaluate both the appearance and disappearance of a solid in a solution or the overall surface of solid particles in a suspension. In the first case, it is possible to determine the nucleation point of a crystallization process (first appearance of a solid) and to measure the solubility point (disappearance of the last crystal).

In lab experiments visual detection of appearance and disappearance of crystals is an usual practice, which is still applied today (Mojumdar *et al.*, 2011; Sahin *et al.*, 2007). More accurate measurements are provided by optical fiber probes (Liang *et al.*, 2004). Turbidimeter is the most widely used instrument to detect the solution cloud and clear points, corresponding to nucleation and solubility points (Moscosa-Santillàn *et al.*, 2000). Avantium (2011) developed a system based on turbidimetric measurements to simultaneously measure the solubility in solutions contained in 16 vials, operating a cooling crystallization. This latter instrument is very helpful to make a quick screening of suitable solvents for a pharmaceutical cooling crystallization process.

Recently, a sensor based on the forward beam reflectance method (FBRM) has been adopted for investigation on nucleation (Schöll *et al.*, 2006; Mitchell *et al.*, 2011); it is accurate as a turbidimeter in the detection of a nucleation point (He *et al.*, 2010), but it is one order of magnitude more expensive.

6.2
Measurement of Nucleation and Solubility Points

As mentioned above, the appearance of nuclei in a bench scale vessel made of glass may be determined by the human eye, but this practice is not feasible in industrial equipment. In this case, in situ sensors are used. Nucleation is detected by turbidimetry in correspondence to a sharp attenuation of the light intensity transmitted through the solution. However, at the onset of nucleation the light

Industrial Crystallization Process Monitoring and Control, First Edition. Edited by
Angelo Chianese and Herman J. M. Kramer.
© 2012 Wiley-VCH Verlag GmbH & Co. KGaA. Published 2012 by Wiley-VCH Verlag GmbH & Co. KGaA.

intensity acquired by the receiver is the sum of the direct attenuated light and the forward scattered light and the sum of the intensities of these two contributions may reduce the measurement resolution.

Turbidimetry is popular, since it also provides the measurement of the suspension density up to a rather high value, but its sensitivity in the detection of the nucleation point has not yet been fully ascertained. Most commercial turbidimeters designed for measuring low turbidities give comparatively good indications of the intensity of light scattered in one particular direction, predominantly at right angles to the incident light. Turbidimeters with scattered-light detectors located at 90° to the incident beam are called *nephelometers*; they are relatively unaffected by small differences in design parameters and therefore are specified as the standard instruments for measurement of low turbidities.

The intensity of light acquired by a nephelometer is only due to the scattered light caused by the appearance of crystals in suspension and high sensitivity with respect to the detection of the nucleation phenomenon is expected.

This work reports a comparison between nucleation measurements made by eye, a turbidimetric sensor, and a nephelometric one with respect to different model systems, exhibiting a very different nucleation rate: potassium sulfate and dextrose monohydrate in aqueous solutions.

6.3
The Developed Turbidimetric and Nephelometric Instruments

Two different instruments were developed at the University of Rome for nucleation detection: a turbidimeter and a nephelometer (Chianese *et al.*, 2005; Chianese, Bravi, and Mascioletti, 2005).

Both instruments consist of a primary sensor, to be put in contact with the solution, and an optical system to provide the light source and the light intensity measurement. This optical measurement system is the same for the two instruments, whereas the primary sensor consists of two glass fibers, 1 mm in diameter, on the opposite side or on the perpendicular side (at 90°) for the turbidimeter and the nephelometer, respectively (see Figure 6.1). The optical system worked as follows. The light source, contained in the electronic block, is a laser diode (wavelength 650 nm), whereas the light intensity measurement device is a digital photodetector which produces a digital signal of 50% duty cycle at a frequency proportional to the brightness of the acquired light intensity. A further digital photodetector is placed beside the laser diode to monitor the emitted light stability.

The fibers that connect the immersion probe to the electronic block are step-index multimode ones with a 600 µm core.

The photodetector has a dynamics of 120 dB which provides a sensibility (<1 NTU) capable of detecting both the nucleation and solubility point with high accuracy.

A microcontroller in the electronic device makes the following actions:

- reading the signals from the digital photodetectors;

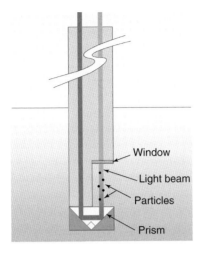 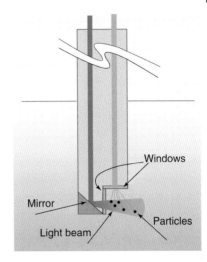

Figure 6.1 Schemes of a turbidimeter and nephelometer.

- handling such signals in order to detect the state of the solution, that is absence or presence of crystals in suspension; and
- sending information to a computer via an RS-232 serial port or to an industrial control panel via a 4–20 mA output port.

The accuracy of the measurement is determined by the sensitivity of the microcontroller with respect to the light flux, equal to $10\,nW/mm^2$.

6.4
The Examined Systems

The developed instruments were used to measure both the nucleation and the solubility point of two different classes of aqueous solutions, containing dextrose monohydrate and potassium sulfate as solutes, respectively. Both these solutes exhibit a direct solubility in water. The used dextrose was supplied by Roquette Italia and had a purity of 99.4% b.w. Impurities consisted mainly of maltose and malto-triose derived from the incomplete hydrolysis of starch. Potassium sulfate was an RPE grade product supplied by Carlo Erba, with a content of impurities less than 0.2%.

The adopted experimental apparatus consisted of a 500 ml jacketed glass vessel connected with a thermostatic bath. The vessel was magnetically stirred at a constant speed of 200 rpm and lighted by a lamp placed behind it, in order to visually detect the nucleation or the dissolution of the crystals. The temperature, measured by a digital thermometer with a precision of $\pm 0.1\,°C$, was continuously recorded.

The measurement of the nucleation point was made by the following procedure. A saturated solution was prepared at a fixed temperature, putting water in contact with an excess of solid and the slurry was stirred in a thermostated vessel for 24 h. This slurry was then filtered and the clear mother liquor was poured into a jacketed stirred vessel, 0.5 l in capacity, operating at the saturation temperature. The temperature was then lowered at a constant rate, equal to 10 °C/h. The nucleation temperature was identified by eye when a cloud of very small crystals inside the solution appeared and by the instruments when a meaningful change of the acquired light intensity was detected. For the runs concerning the dextrose solutions at the beginning of each experiment the solution was seeded with four to five big crystals (>425 μm obtained by sieving) to induce nucleation. In this case the secondary nucleation point by catalytic mechanism was measured.

The measurement of the solubility point was made on the basis of the detection of the temperature at which the dissolution of the last crystal occurs. The measurement procedure consists of firstly causing the nucleation process by cooling and then inverting the temperature change, that is the suspension was heated up, in order to progressively dissolve the fine crystals. The equilibrium temperature was detected when the last crystal dissolved.

The adopted procedure to measure the solubility point is a little different from the usual one, consisting of seeding a few big crystals in a supersaturated solution and inducing their dissolution by a continuous solution heating. The proposed procedure is more accurate than the traditional one since the dissolution of fines, slightly grown from the nuclei, is very rapid in correspondence to the solubility point.

6.5
Obtained Results

The first series of experiments concern the measurement of the nucleation point determined by eye, turbidimetry, and nephelometry for the two examined aqueous solutions systems. For each system, two series of experiments were performed by carrying out three runs under constant operating conditions. The obtained results for potassium sulfate are reported in Table 6.1. The temperature measurement range concerns data reproducibility.

From the data in Table 6.1 we may observe that: (i) the nucleation temperature detected by nephelometry is higher than those detected by turbidimetry and

Table 6.1 Primary nucleation points for aqueous solutions of potassium sulfate.

	Temperature of nucleation (°C)		
$c_{K_2SO_4}$ (g/100 g water)	Visual method	Nephelometry	Turbidimetry
14.0	24.6 ± 0.8	26.0 ± 0.3	25.0 ± 0.6
15.7	34.6 ± 0.6	35.2 ± 0.2	35.3 ± 0.4

by eye, thus indicating that nephelometry is the most sensitive method; (ii) the reproducibility of the measurement is again the best one in the case of nephelometry, since the experimental data measured under the same conditions exhibit the smallest deviations.

Some experiments were then performed with dextrose monohydrate solutions. The nucleation rate for this system is very low. In particular, when the experiments are performed in the absence of seeding the nucleation phenomenon is stochastic; thus, the nucleation point may occur in a wide temperature range. Moreover, the first appearance of nuclei gives rise to a very weak cloud of crystals and the detection of nucleation is difficult. For these reasons only runs with seed addition were performed and the secondary nucleation temperature was measured.

The results obtained for this system are reported in Table 6.2.

Again, the detection of the nucleation point by nephelometry takes place earlier with respect to turbidimetry and the measurements are more reproducible. However, in this case the visual measurement and the nephelometric one compare quite well each other, whereas by the turbidimetric method a later detection of the nucleation point is obtained. A further series of experiments on secondary nucleation were then carried out by comparing the visual method with the nephelometric one.

Table 6.2 Secondary nucleation points for aqueous solutions of dextrose monohydrate.

	Temperature of nucleation (°C)		
c_{DX} (g/100 g water)	Visual method	Nephelometry	Turbidimetry
63.87	28.5 ± 0.5	28.7 ± 0.2	27.5 ± 0.5
57.41	18.2 ± 0.6	18.5 ± 0.3	18.0 ± 0.5

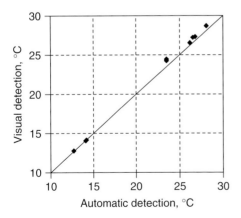

Figure 6.2 Comparison between DX nucleation points by visual and nephelometric (automatic) methods.

The results reported in Figure 6.2 show good agreement between the temperature values given by the two methods.

The use of a sensor also gives the chance of continuously measuring the magma of the suspension after the first crystal appearance. This information may concern either the increase of the magma density due to the growth of the nuclei or the dissolution of the fine crystals present in the suspension, when the solution becomes more or less supersaturated, respectively. Since the mass of crystals generated just after nucleation or disappearing just before the clear point is quite low, the change of the mass suspension may be accurately detected by nephelometry.

An example of the mass density recording along an experiment consisting of a controlled cooling of a saturated solution, followed by a two-step controlled heating of the same solution, is reported in Figure 6.3. The examined system was a potassium sulfate aqueous solution with a concentration of 14 g of salt per 100 g of water. The solution, initially saturated, became progressively supersaturated by cooling at a rate of 10 °C/h. After the detection of the nucleation point at around 26 °C the solution was furtherly cooled by 1 °C in order to make evidence of the increase of the magma density due to both heterogeneous nucleation and nuclei growth. The solution temperature was then increased in order to induce crystal dissolution at a heating rate of 10 °C/h. The light intensity detected by the instrument along this period of time firstly reached a maximum and then progressively decreased. Just before the expected solubility point the heating rate was decreased to 6 °C/h in order to slightly dissolve the residual crystals. The same

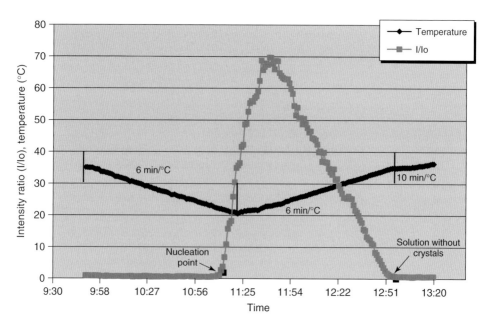

Figure 6.3 Detection of the nucleation and solubility point by a nephelometer.

transmitted light intensity of the saturated solution was measured by the instrument when the last crystal was dissolved, thus showing the attainment of the solubility point, equal to 30 °C.

References

Chianese, A., Bravi, M., Cugola, M., and Mascioletti, A. (2005) Torbidimetri di processo per il controllo di qualità nella cristallizzazione industriale. IV Congress on Metrology and Quality Proceedings, Vol. 2, pp. 93–96.

Chianese, A., Bravi, M., and Mascioletti, A. (2005) Metodo e strumento optoelettronico per la misura della solubilità di sostanze chimiche. Italian Patent no. RM2005A000182.

He, G., Tjahjono, M., Shan Chow, P., Tan, R.B.H., and Garland, M. (2010) In situ determination of metastable zone width using dielectric constant measurement. *Org. Process Res. Dev.*, **14**, 1469–1472.

Liang, K., White, G., and Wilkinson, D. (2004) An examination into the effect of stirrer material and agitation rate on the nucleation of L-glutamic acid batch crystallized from supersaturated aqueous solutions. *Cryst. Growth Des.*, **4** (5), 1039–1044.

Mitchell, N.A., Frawley, P.J., and Ó'Ciardhá, C.T. (2011) Nucleation kinetics of paracetamolethanol solutions from induction time experiments using Lasentec FBRM. *J. Cryst. Growth*, **321** (1), 91–99.

Mojumdar, S.C., Madhurambal, G., Mariappan, M., and Ravindran, B. (2011) Nucleation kinetics of a new nonlinear optical crystal – urea-thiourea zinc chloride. *J. Therm. Anal. Calorim.*, **104** (3), 901–907.

Moscosa-Santillàn, M., Bals, O., Fauduet, H. Porte, C., and Delacroix, A. (2000) Study of batch crystallization and determination of an alternative temperature-time profile by on-line turbidity analysis application to glycine crystallization. *Chem. Eng. Sci.*, **55**, 3759–3770.

Sahin, O., Dolas, H., and Demir, H. (2007) Determination of nucleation kinetics of potassium tetraborate tetrahydrate. *Cryst. Res. Technol.*, **42** (8), 766–772.

Schöll, J., Vicum, L., Müller, M., and Mazzotti, M. (2006) Precipitation of L-glutamic acid: determination of nucleation kinetics. *Chem. Eng. Technol.*, **29** (2), 257–264.

www.avantium.com/pharma/ accessed (2011).

7
Speed of Sound

Joachim Ulrich and Matthew J. Jones

7.1
Introduction

While the majority of in-line measurement techniques currently in use probe either a property of the solution or depend on the presence of solids in a slurry/solution, the speed of sound depends on solution properties and also changes in the presence of solids. It is, therefore, one of only few techniques that is suitable for in-process monitoring of solute concentration as well as nucleation.

In a given medium, sound travels with a velocity v, which depends on the density ρ and the adiabatic compressibility c_{ad} of the medium (Equation (7.1)). Both density and adiabatic compressibility depend on temperature, concentration, and pressure. For liquids, the pressure dependence can be neglected, as here only pressures higher than 100 bar lead to significant changes in density and adiabatic compressibility. For gaseous phases, the pressure dependence cannot be neglected:

$$v = \sqrt{\frac{\rho}{c_{ad}}} \qquad (7.1)$$

As a consequence, it is possible *to map temperature and concentration fields against the speed of sound*, providing that it is possible to detect changes in the velocity of sound with respect to concentration and temperature with sufficient accuracy.

7.2
In-Process Ultrasound Measurement

The general setup for the measurement of sound velocity in solutions and melts is shown in Figure 7.1. The components are commercially available and the setup consists of the ultrasound sensor, a controller, and a computer for data acquisition. The ultrasound sensor emits a low-frequency ultrasonic wave. Due to the low frequency, the energy input is not sufficient to cause cavitations and therefore the ultrasound probe neither influences nor promotes nucleation from the solution. The application of sound velocity measurement for the purpose of

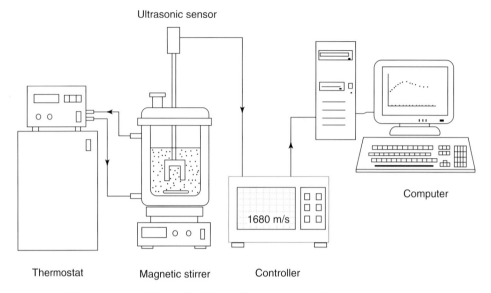

Figure 7.1 General setup for ultrasound measurement of concentration with respect to supersaturation.

concentration measurement was patented (Ulrich and Glade, 1999) and described by the group of authors (Omar, Strege, and Ulrich, 1999; Omar, 1999; Omar and Ulrich, 1999a, b; Sayan and Ulrich, 2002; Garside *et al* 1982).

The sound velocity is measured by recording the time required for the sound to traverse the known distance between the sound generator and the receiver. The sound velocity is determined by the properties of the medium between the emitter and the receiver and by the temperature (via the temperature dependence of the density and adiabatic compressibility). Figures 7.2 and 7.3 show examples of the concentration dependence of the sound velocity against temperature (Omar, Strege, and Ulrich, 1999) and temperature and concentration (Strege, 2004).

7.3
Determining Solubility and Metastable Zone Width

A successful determination of the state of a given system is contingent upon the following two conditions:

- The difference in sound velocity of two different states has to be sufficiently large to discriminate between one state and the other.
- The measurement time correlates with the concentration of only one component in the system.

Both of the above conditions may be affected by the change of density and adiabatic compressibility accompanying any change in the system composition

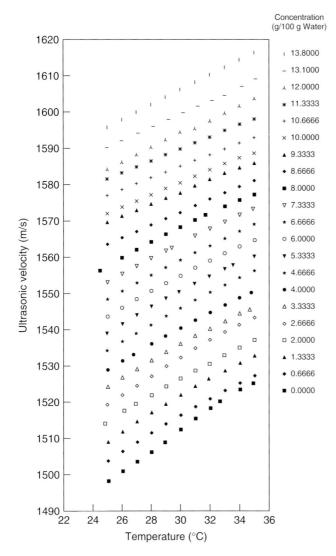

Figure 7.2 Experimentally determined dependence of sound velocity upon temperature for various concentrations of potassium sulfate in water (Omar, Strege, and Ulrich, 1999; Omar and Ulrich, 1999a).

and temperature. In some cases, changes in ρ and c_{ad} have opposing slopes, thus canceling the effect on the sound velocity. In that case no dependence of sound velocity on concentration is observed, even if the concentration change is significant. As a consequence, the method will not provide a useful signal and therefore fails. For this reason, the applicability of the technique to the given problem has to be established on a case-to-case basis.

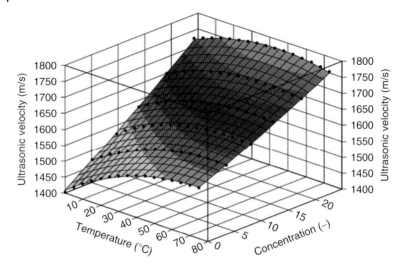

Figure 7.3 Ultrasonic velocity for magnesium sulfate solutions with a concentration up to 25 wt% (Strege, 2004).

A range of materials, both organic and inorganic, are known, where sound velocity measurement can be gainfully employed for concentration measurements. Among these are:

Compound	Solvent	Author
Fluoranthene	Trichlorethylene	Marciniak (2002)
$NiSO_4 \cdot 6H_2O$	H_2O	Mohnicke, Strege, and Ulrich (2002)
Ketogulonic acid	H_2O	Mohnicke, Strege, and Ulrich (2002); Benecke, Bay, and Ulrich (2002)
$C_6H_8O_7 \cdot H_2O$	H_2O	Omar, Strege, and Ulrich (1999)
$CH_3COONa \cdot 3H_2O$	H_2O	Omar, Strege, and Ulrich (1999)
K_2SO_4	H_2O	Omar, Strege, and Ulrich (1999)
KCl	H_2O	Omar, Strege, and Ulrich (1999)
Oxalic acid	H_2O	Omar and Ulrich (2006)
$MgSO_4 \cdot 6H_2O$	H_2O	Strege (2004)
$MgSO_4 \cdot 7H_2O$	H_2O	Strege (2004)
$ZnSO_4 \cdot 7H_2O$	H_2O	Strege (2004)
$C_9H_{11}NO_2$ or $C_6H_5CH_2CH(NH_2)CO_2H$	–	Strege (2004)
$Na_2B_4O_7 \cdot 10H_2O$	H_2O	Gürbüz and Özdemir (2003), Strege (2004)
$CaCl_2 \cdot 2H_2O$	H_2O	Strege (2004)
$C_{18}H_{36}O_2$	Methanol	Strege (2004)
$(NH_4)_2SO_4$	H_2O	Titiz-Sargut and Ulrich (2003)
$KAl(SO_4)_2 \cdot 12H_2O$	H_2O	Titiz-Sargut and Ulrich (2002)

Some materials known not to be amenable to sound velocity measurements are, for example, sugars and NaCl.

Since solubility, metastable zone width, and crystal growth rates are functions of concentration, these can be quantified using the technique. How this can be achieved is discussed in the following using the example of ammonium sulfate.

The temperature of an undersaturated solution is changed with a constant rate. In the case of $(NH_4)_2SO_4$ (Titiz-Sargut and Ulrich, 2003), where the slope of the solubility curve is positive, the temperature is decreased in order to increase the level of saturation. For systems where the solubility has a negative slope with respect to temperature, the latter has to be increased in order to increase the level of saturation. In the following, a positive slope of the solubility curve is implicitly assumed. As long as the solution remains homogeneous, the sound velocity is a monotonic function of temperature and therefore possesses no discontinuities (Figure 7.4). As soon as a *nucleation* occurs, the concentration of the solute in the homogeneous liquid phase decreases instantaneously. The solid phase now present has a different density than the solution and therefore the sound velocity exhibits a discontinuity. If, at this point, the temperature gradient is reversed (in this case the solution is now heated with the same, constant rate), the sound velocity now increases with increasing temperature. The difference between the values on cooling and on heating is a consequence of the different nature of the system in the cooling regime (homogeneous solution) and the heating regime (suspension of solid in liquid). The two curves naturally meet at the *saturation point*, as here the solid generated by the nucleation event on cooling has completely dissolved. The temperature difference between the nucleation point and the saturation point represents the width of the metastable zone.

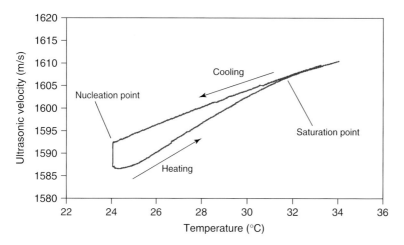

Figure 7.4 Determination of the saturation point and the metastable zone width of a solution or a melt (here: aqueous K_2SO_4) (Ulrich and Omar, 1999).

Conducting this experiment at different concentrations results in both the saturation curve as well as a curve representing the upper limit of the metastable zone. Employing different cooling (heating) rates provides access to the rate dependence of the metastable zone.

At this point, a cautionary note is necessary. As shown by Holmes, Challis, and Wedlock (1993), the ultrasound velocity can be severely attenuated in the presence of high particle concentrations, that is, for large suspension densities. A simple remedy is to isolate the ultrasonic probe from solids, but not the solution, by means of a porous mesh. This approach was used in Titiz-Sargut and Ulrich (2003) and a comparison between measurements both with and without the protective mesh was presented.

Figure 7.5 shows how a solubility curve can be determined together with the metastable zone width by measuring the temperature dependence of the sound velocity for a range of concentrations. A similar figure can be found, for example, for $KAl(SO_4)_2 \cdot 12\,H_2O$ as reported by Titiz-Sargut and Ulrich (2002).

As the solubility is a thermodynamic property of the system, it is not time dependent. The *metastable zone width*, however, is not a thermodynamically defined quantity and depends on the process conditions and can be influenced by temperature (see, Omar and Ulrich, 1997; Titiz-Sargut and Ulrich, 2002 for KCl or $KAl(SO_4)\cdot 12H_2O$, respectively) and cooling rate and the presence of impurities (see e.g., eight additives and their influence on magnesium sulfate grown from aqueous solutions, Strege, 2004, or four additives and their influence on $KAl(SO_4)_2 \cdot 12\,H_2O$, Titiz-Sargut and Ulrich, 2002). Figures 7.6 and 7.7 show examples for each of these cases.

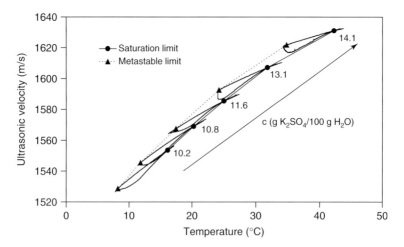

Figure 7.5 A solubility curve can be established by measuring the sound velocity for a range of concentrations. Employing the procedure outlined in the text also provided the (concentration dependent) metastable zone width (Strege, Omar, and Ulrich, 1999).

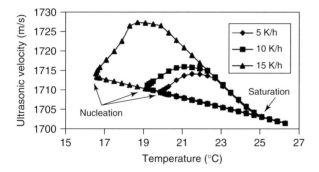

Figure 7.6 Influence of the cooling rate on the width of the metastable zone of KCl (Omar and Ulrich, 1997; Ulrich and Strege, 2002).

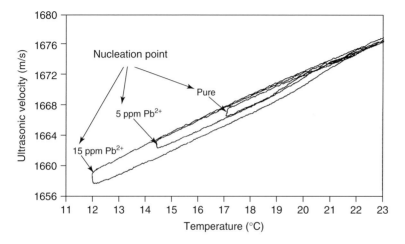

Figure 7.7 Influence of the presence of additives (here Pb^{2+}) on the width of the metastable zone of KCl at a cooling rate of 4 K/h (Ulrich and Omar, 1999; Ulrich and Strege, 2002).

7.4
Measuring Crystal Growth Rates

Crystal growth rates can be determined by *recording desupersaturation curves*, that is, by continuous measurement of the change of supersaturation while crystals are growing as a result of the initially induced supersaturation. Growth rates are obtained from a single experiment by back calculation from the concentration change deduced from the change in sound velocity. A method for calculating growth rates using desupersaturation curves is described by Chivate and Tavare (1975) and Tavare and Chivate (1978) as well as by Omar, Strege, Garside *et al.* (1982) and Ulrich (1999a). However, measuring the change of supersaturation by means of

7.5
Detecting Phase Transitions with Ultrasound

Solvent-mediated phase transitions, such as solvate–solvate or solvate–desolvate transformations, can also be detected by ultrasound. In these cases, the composition of the solution changes, which may result in a change of sound velocity

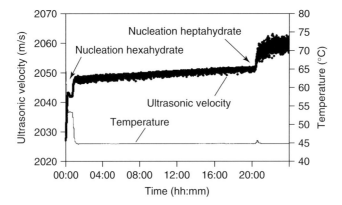

Figure 7.8 Phase transformation of $MgSO_4 \cdot 6H_2O$ to $MgSO_4 \cdot 7H_2O$ as evidenced by the change of the ultrasound velocity (Strege, 2004).

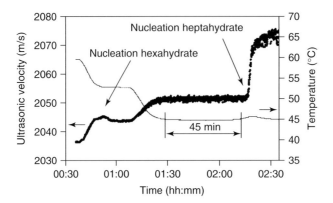

Figure 7.9 Phase transformation of magnesium sulfate in the presence of 0.5 wt% KCl (Strege, 2004).

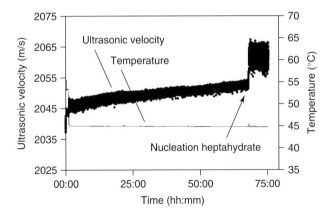

Figure 7.10 Phase transformation of magnesium sulfate in the presence of 0.5 wt% borax (Strege, 2004).

in the medium. The work of Strege (2004) clearly demonstrated the value of applying this technique in studying such phase transitions. Figures 7.8–7.10 show measurements of sound velocity during the transformation of $MgSO_4 \cdot 6H_2O$ to $MgSO_4 \cdot 7H_2O$ in a pure solution (Figure 7.8) and in the presence of additives (Figures 7.9 and 7.10). The transition temperature is 48 °C.

In these phase-transformation experiments, solutions with a saturation temperature of 56 °C are cooled to 53 °C. If nucleation of the hexahydrate is not observed within 15 min, a seed crystal is added in order to induce nucleation. After depletion of supersaturation by nucleation and crystal growth, the suspension is cooled to the final transformation temperature below the transition point. In this series of experiments, the solution-mediated phase transformation was allowed to proceed at a constant temperature of 45 °C. After reaching the final temperature, the ultrasound velocity increases linearly with time up to the point where nucleation of the stable phase begins. The starting point for nucleation of the heptahydrate is indicated by a distinct increase in ultrasound velocity and by a peak in the temperature measurement. In the pure solution, nucleation of the stable phase can be observed after an induction period of ∼19 h, as can be seen in Figure 7.8.

Two additives, potassium chloride and borax, were also investigated. Both influence the induction time for nucleation of the heptahydrate significantly. The effect of KCl on the phase transformation is shown in Figure 7.9.

Experiments were carried out in the same manner as for pure solutions. Adding KCl clearly accelerates the transformation. In solutions with 0.5 wt% KCl, nucleation of the heptahydrate starts after a much reduced induction time of 45 min. In contrast, transformation in the presence of 0.5 wt% borax is distinctly delayed (Figure 7.10). Nucleation of the stable phase can only be observed after an induction accompanied by a distinct increase of the ultrasound velocity and an increase in temperature.

References

Benecke, I., Bay, K., and Ulrich, J. (2002) Determination of the metastable zone width and control of batch crystallizer by ultrasound technique, in *Proceedings of the 15th International Symposium on Industrial Crystallization, 15th–18th September, Sorrento, Italy* (ed. A. Chianese), Chemical Engineering Transactions, AIDIC, Milano, pp. 503–507.

Chivate, M.R. and Tavare, N.S. (1975) Growth rate measurement in DTB crystallizer. *Chem. Eng. Sci.*, **30**, 354–355.

Garside, J., Gibilaroand, L.G., and Tavare, N.S. (1982) Evaluation of crystal growth kinetics from a desupersaturation curve using initial derivations. *Chem. Eng. Sci.*, **37**, 1625–1628.

Gürbüz, H. and Özdemir, B. (2003) Experimental determination of the metastable zone width of borax decahydrate by ultrasonic velocity measurement. *J. Cryst. Growth*, **252**, 343–349.

Holmes, K.A., Challis, E.R., and Wedlock, D.A. (1993) A wide bandwidth study of ultrasound velocity and attenuation in suspension: comparison of theory with experimental measurement. *J. Colloid Interf. Sci.*, **156**, 261–268.

Marciniak, B. (2002) Density and ultrasonic velocity of undersaturated and supersaturated solutions of fluoranthene in trichloroethylene, and study of their metastable zone width. *J. Cryst. Growth*, **236**, 347–356.

Mohnicke, M., Strege, C., and Ulrich, J. (2002) The application of ultrasonic sensors for the control of crystallization processes – options and limitations, in *Proceedings of the 9th International Workshop on Industrial Crystallization (BIWIC9), 11–12th September, Halle (Saale), Germany* (ed. J. Ulrich), Martin-Luther-Universität Halle-Wittenberg, Halle (Saale), pp. 80–87.

Omar, W. (1999) Zur Bestimmung von Kristallisationskinetiken auch unter der Einwirkung von Additiven mittels Ultraschallmesstechnik. PhD thesis. Universität Bremen, Germany.

Omar, W., Strege, C., and Ulrich, J. (1999) Bestimmung der Breite des metastabilen Bereichs von realen Lösungen mittels der Ultraschallmesstechnik zur on-line-Prozessregelung bei Kristallisationsverfahren. *Chem. Tech.*, **51** (5), 286–290.

Omar, W. and Ulrich, J. (1997) Application of ultrasonics in the control of crystallization processes, in *Crystal Growth of Organic Materials 4* (ed. J. Ulrich), Verlag Shaker, Aachen, pp. 294–301, Proceedings, 17th–19th August, Bremen, Germany.

Omar, W. and Ulrich, J. (1999a) Application of ultrasonics in the on-line determination of supersaturation. *Cryst. Res. Technol.*, **34** (3), 379–389.

Omar, W. and Ulrich, J. (1999b) Ultrasonic methods for the control and study of batch crystallization processes, in *Proceedings of the 14th International Symposium on Industrial Crystallization, 12th-16th September, Cambridge, UK* (eds J. Garside, R.J. Davey, and M. Hounslow), IChemE Rugby, Cambridge.

Omar, W. and Ulrich, J. (2006) Solid liquid equilibrium, metastable zone and nucleation parameters of the oxalic acid–water system. *Cryst. Growth Des.*, **6** (8), 1927–1930.

Sayan, P. and Ulrich, J. (2002) The effect of particle size and suspension density on the measurement of ultrasonic velocity in aqueous solutions. *Chem. Eng. Process.*, **41**, 281–287.

Strege, C. (2004) On (Pseudo-) Polymorphic Phase Transformations. PhD thesis. Martin-Luther-Universität Halle-Wittenberg, Online publication: http://sundoc.bibliothek.uni-halle.de/diss-online/04/04H318/index.htm.

Strege, C., Omar, W., and Ulrich, J. (1999) Measurement of metastable zone width by means of ultrasonic devices, in *Proceedings of the 7th International Workshop on Industrial Crystallization (BIWIC7), 6th–7th September, Halle (Saale), Germany* (ed. J. Ulrich), Martin-Luther-Universität Halle-Wittenberg, Halle (Saale), pp. 219–230.

Tavare, N.S. and Chivate, M.R. (1978) Growth rate correlation for potassium sulfate crystals in a fluidized bed crystallizer. *Chem. Eng. Sci.*, **33**, 1290–1292.

Titiz-Sargut, S. and Ulrich, J. (2002) Influence of additives on the width of the metastable zone. *Cryst. Growth Des.*, **2** (5), 371–374.

Titiz-Sargut, S. and Ulrich, J. (2003) Application of a protected ultrasound sensor for the determination of the width of the metastable zone. *Chem. Eng. Process.*, **42**, 841–846.

Ulrich, J. and Glade, H. (1999) Perspectives in control of industrial crystallizers, in *International Workshop "Modelling and Control of Industrial Crystallization Processes" Proceedings*, IBM Centre, La Hulpe.

Ulrich, J. and Omar, W. (1999) Verfahren zur Ableitung von Parametern für die Prozessführung sowie zur Online-Überwachung des Kristallisationsvorganges und dazu verwendbare Anordnung. Offenlegungsschrift DE 19741667 A1, Deutsches Patent- und Markenamt München, Germany.

Ulrich, J. and Strege, C. (2002) Some aspects of the importance of metastable zone width and nucleation in industrial crystallizers. *J. Cryst. Growth*, **237–239**, 2130–2135.

8
In-Line Process Refractometer for Concentration Measurement in Sugar Crystallizers

Klas Myréen

8.1
Introduction

One of the more accurate ways to estimate a solute concentration is based on the measurement of the refractive index (RI) of the solution. This chapter describes the principle of an in-line refractometer and its use in sugar industry.

It was in 1874, when the German scientist Prof. Ernst Abbé was credited the innovation of the refractometer. Since then refractometers have been a common laboratory instrument to measure concentrations of solutions. For decades refractometers have been in use in sugar manufacturer's laboratories for accurate and reliable Brix measurement (measurement of amount of sucrose in water).

There were several attempts made in the past to apply Abbé's technology to an instrument that could be used for reliable in-line monitoring of concentrations.

In the 1970s, the first in-line process refractometers were introduced to the market. Later there was introduction of the digital process refractometers (DPRs), which does not have any drift of the measurement and is an easy and effective way to measure the concentration of solute in solutions.

The refractometer determines the change of speed of light when light travels from one media to another. When the light passes from one media to another, for example, air to water, the bending angle will change (see Figure 8.1). The *refractive index* (symbolized as n_D) is defined as the speed of light in air divided by the speed of light in the medium according to Snell's law, which states that the ratio of the sines of the angles of incidence and refraction is equivalent to the ratio of phase velocities in the two media, or equivalent to the opposite ratio of the indices of refraction (Groetsch, 1996; Clevett, 1986).

At present, in-line process refractometers have been proved to be accurate and reliable instruments for the sugar concentration measurements and measurement of other solute concentration as well. The RI measured by a DPR is not affected by the crystals and bubbles in the crystallizer and can be therefore properly used to estimate the mother liquor concentration (the dissolved matter) in the vacuum pan application. The mother liquor concentration signal can be used as an input parameter for the calculation of the supersaturation (SS) value in the vacuum pan.

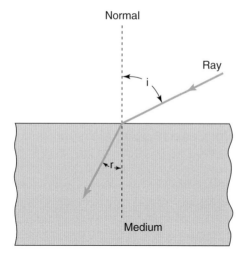

Figure 8.1 Refraction of light (i: angle of incidence. r: angle of reflection).

8.2
Measurement Principle

There are few ways to detect the RI of the solution. The most typical way, and most used in process industry, is to determine the critical angle by the total internal reflection according to Snell's law. The critical angle of refraction is utilized by using a prism in contact with the liquid, a light emitting diode (LED) light source, and a detector.

The LED wavelength commonly in use for DPR is a visible yellow light, equal to 580 nm, that is, the same as laboratory electronic refractometers.

The light from the light source (L) in Figure 8.2 is directed to the interface between the prism (P) and the process medium (S). Two of the prism surfaces (M) act as mirrors bending the light rays so that they meet the interface at different angles.

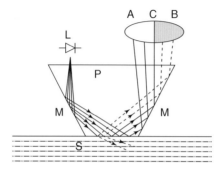

Figure 8.2 Critical angle refractometer principle.

The reflected rays of light form an image (ACB), where (C) is the position of the critical angle ray. The rays at (A) are totally internally reflected at the process interface, the rays at (B) are partially reflected and partially refracted into the process solution. In this way, the optical image is divided into a light area (A) and a dark area (B). The position of the shadow edge (C) indicates the value of the critical angle. The refractive index n_D can then be determined from this position.

The refractive index n_D changes with the process solution concentration and temperature. When the concentration increases, the RI normally increases. At higher temperatures, the RI is smaller than at lower temperatures. The shadow edge moves in the optical image depending on the concentration and the temperature, as shown in Figure 8.3. The color of the solution, gas bubbles, or undissolved particles do not affect the position of the shadow edge (C).

The position of the shadow edge is measured digitally using a CCD element (Figure 8.4) and is converted to an RI value n_D by a processor inside the sensor. This value is then transmitted together with the process temperature to a processor unit, for further processing, display, and transmission. The processor contains the correlation curve to convert the RI and temperature to a concentration value (Brix, concentration by weight, or direct RI) for the examined media.

The use of the CCD element and monitoring the shadow edge mean that the measurement is drift free and does not change with time.

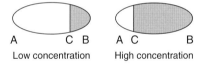

Figure 8.3 Optical images and position of shadow edge (C).

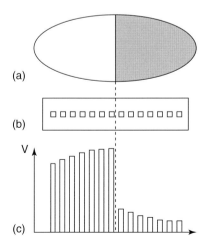

Figure 8.4 Optical image detection. (a) Optical image; (b) CCD element; (c) CCD output.

8.3
In-Line Instrument Features and Benefits

8.3.1
Accuracy

In-line refractometers measure with the same accuracy as a laboratory refractometer, that means about RI accuracy of ±0.0001, which corresponds on the sugar concentration curve to ±0.05 Brix (% b/w). Higher accuracies and repeatability's are possible with special arrangements. The measurement range is typically preset to 1.32–1.53, which corresponds to 0–100 Brix.

8.3.2
Concentration Determination

Generally solute concentration is a nonlinear function of the critical angle and the refractive index n_D.

Figure 8.5 shows the concentration in weight percentage of a sucrose solution (Brix scale) as a function of the RI. The microprocessor in the electronic assembly performs complete linearization, temperature compensation, and various functional tests.

The calibration curve is used in the instrument to calculate the solute concentration from the RI value. The calibration curve varies depending on the media to be measured. The magnitude of process temperature compensation on the concentration signal depends on the media to be measured and variations of process temperatures.

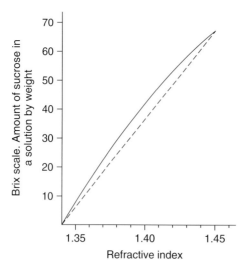

Figure 8.5 Correlation of refractive index to sucrose by weight in solution (Brix scale).

8.3.3
Process Temperature Compensation Factor

In general, the RI of a substance decreases with increasing temperature and vice versa. Due to this, the in-line refractometer has a built-in temperature compensation factor.

The temperature compensation factor depends on the examined media. Typically for a sugar solution, an increase of temperature of 1 °C decreases the RI by 0.0002 units.

When the temperature compensation for a solution has been determined, it is used to compensate for process temperature variations and the output concentration signal of the instrument is temperature compensated, fully automated.

For automated temperature compensation, the sensor tip contains a process temperature probe. The microprocessor calculates the concentration based on the temperature data, RI data, and the calibration curve.

8.3.4
Process Sensor

Figure 8.6 shows a typical cut-away picture of an in-line process refractometer sensor.

The measurement prism (A) is flush mounted to the surface of the probe tip. The prism (A) and all the other optical components are fixed to the solid core module (C), which is spring loaded (D) against the prism gasket (B). The light source (L) is

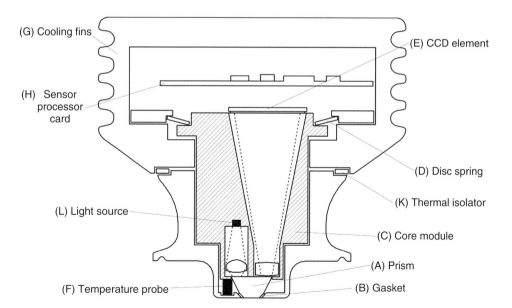

Figure 8.6 Process sensor cut-away.

Figure 8.7 Probe process refractometer.

a yellow LED, and the receiver is a CCD element (E). The electronics is protected against the process heat by a thermal isolator (K) and cooling fins (G).

The sensor processor card (H) receives the raw data from the CCD element (E) and the Pt-1000 process temperature probe (F) and then calculates the refractive index n_D and the process temperature T. This information is transmitted to the processor, which is in an indicating transmitter box or integrated to sensor.

For insertion in sugar vacuum pan, a probe sensor is recommended (see Figure 8.7).

The concentration value is displayed and can be taken out as a 4–20 mA and Ethernet UDP/IP signal.

8.4
Features and Benefits

A process refractometer with an inside conversion of the digital signal provides a drift free, no maintenance, and precalibrated measurement. It is an easy instrument to install and does not require any maintenance.

A critical angle refractometer has also the benefit of measurement in black, turbid, or dark liquids. It just detects the shadow edge between light and dark and light intensity is not affected by the concentration value.

If the prism is mounted in an angle against the process flow, a self-cleaning effect makes the use of cleaning systems unnecessary in many cases. Instruments installed in high-purity sugar crystallization do not typically require a cleaning system. The prism is kept clean by the abrasive effect caused by the growing crystals in the pan.

For a sugar vacuum pan, it is always recommended to install the sensor in the pan itself and not in a by-pass pipe.

8.5
Example of Application in the Crystallization

Refractometer is used extensively in sugar crystallization, but it can be applied for other types of crystallization processes as well, whenever the measurement of a solute concentration is needed. Here after an example of application of refractometer measurements for monitoring and control, a sugar vacuum pan is reported (Rózsa, 1996, 2008a, b).

8.5.1
Seeding Point and Supersaturation Control in Sugar Vacuum Pan

Supersaturation and crystal content play a decisive role in sugar crystallization. The knowledge of SS is very important during the full duration of the boiling process. There is no direct method to measure the SS but it can be calculated using the following variables:

- composition of the mother liquor;
- temperature of the mother liquor; and
- solubility of the examined solute in the mother solution.

Obviously, the mother liquor composition and the solubility function have to be separately determined.

It has been proven that the mother liquor composition has by far the strongest effect on SS, so its evaluation is of paramount importance.

During the crystal seeding at SS conditions, the mother liquor concentration greatly changes, and after seeding the growing amount of crystals in suspension have no influence on the RI values.

When SS is calculated from measurements of primary data like temperature, density, conductivity, viscosity, or RF, reproducible results can only be obtained if the mother liquor composition does not change. It can be argued if this is ever the case in practical sugar boiling.

Such a problem is overcome with the use of a process refractometer, which provides reliable information on the mother liquor concentration, at the seeding point and after the seeding has been done, during the decrease of SS (Figure 8.8).

The SS value is usually expressed as SS ratio, that is, the sugar actual concentration divided by the saturation one.

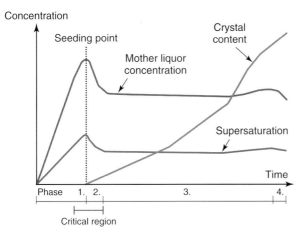

Figure 8.8 Curves of supersaturation and mother liquor Brix.

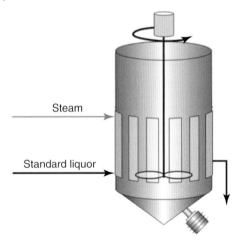

Figure 8.9 DPR installation in a vacuum pan.

In sugar crystallization, the process refractometer is installed directly in the pan (Figure 8.9). The probe is swept by the flow in the pan and the rubbing effect, which is caused by the growing crystals keeps the prism clean.

The control of the SS by means of a process refractometer, in particular during and shortly after seeding, results in improved crystal size distribution (CSD), less amount of false grain and conglomerates, as well as better results for molasses separation in the centrifugals.

By doing the seeding based on a certain SS value instead of using only primary data, it will result in improved crystal quality and better control and understanding

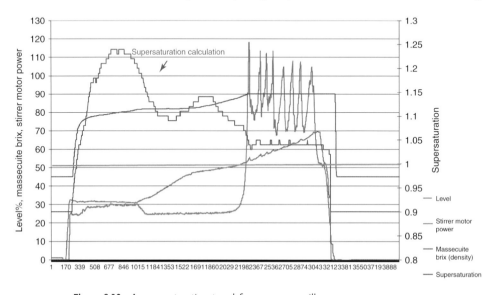

Figure 8.10 A supersaturation trend from a sugar mill.

of the crystallization. The same applies for the time after the seeding has been made.

Instead of traditional manual addition of the seeds, automatic seeding can be done at the determined SS value.

An example of an SS trend from a full seeding sugar vacuum pan can be seen in Figure 8.10, where, apart from the SS value, the crystal content, mother liquor purity, density, massecuite solids content, mean crystal size, and consistency is known either based on the primary data measurement signal or is calculated from those.

By looking at the trend, one can see that in this particular case the SS level after seeding is not constant throughout the batch, as needed for optimum performances, for the absence of a feedback control.

8.6
Conclusion

Usage of an *in-situ* RI instrument for concentration measurement is feasible and allows a continuous monitoring of SS in an industrial crystallizer. The reason for this is the measurement principle which is not affected from bubbles, particles, crystals, or color. The refractometer here described can withstand the process conditions usually applied in industrial crystallizers; thus, it can be adopted for in-line process monitoring. The achieved RI measurements in combination with a preliminary calibration give rise to accurate values of the solute concentration.

The RI is a worthwhile property to measure the mother liquor solute concentration and can thus be used for calculation of SS during sugar vacuum pan boiling. The monitoring of SS real time gives opportunity to do vacuum pan control based on SS instead of traditional ways, thus allowing improved crystallization process performances and better product quality.

References

Clevett, K.J. (1986) Measurement of refractive index, in *Process Analyser Technology*, Chapter 17, John Wiley & Sons, Inc., New York, pp. 707–725.

Groetsch, J. (1996) Refractive index analysers, in *Analytical Instrumentation*, Chapter 14, Instrument Society of America, Research Triangle Park, NC, pp. 269–282.

Rózsa, L. (1996) On-line monitoring of supersaturation in sugar crystallisation. *Int. Sugar J.*, **98** (1176), 660–675.

Rózsa, L. (2008a) Sugar crystallization: look for the devil in the details – part 2. *Int. Sugar J.*, **110**, 729–739.

Rózsa, L. (2008b) Sugar crystallization: look for the devil in the details – part 1. *Int. Sugar J.*, **110**, 403–413.

9
ATR-FTIR Spectroscopy

Christian Lindenberg, Jeroen Cornel, Jochen Schöll, and Marco Mazzotti

9.1
Introduction

Attenuated total reflectance Fourier transform infrared (ATR-FTIR) spectroscopy has been applied successfully to monitor the liquid phase concentration during crystallization processes (Groen and Roberts, 2001; Lewiner *et al.*, 2001; Schöll *et al.*, 2006a, 2007a). IR spectroscopy is used to determine energy differences between vibrational states of molecules in the solid, liquid, and gaseous phase. ATR-FTIR spectroscopy is based on absorption in the mid-infrared region, that is, a photon of infrared radiation of frequency ν_S is absorbed and the molecule is promoted to a higher vibrational state (Figure 9.1). For this absorption process to occur, the energy of the photon must match the separation of vibrational states in the sample (Schrader, 1995; Shriver and Atkins, 1999). The associated absorbance spectrum corresponds to the particular IR wavelengths, which are absorbed by the sample, thus revealing details about its molecular structure.

In ATR spectroscopy, the measuring beam is reflected internally at the interface between an auxiliary medium and the sample. This auxiliary medium must be infrared transparent and of high refractive index (Schrader, 1995). Since the penetration depth is only a few micrometers, typically in the order of the wavelength of the light and depending on the refractive indices, that is, 0.5–5 μm, the ATR technology can be used to measure exclusively the liquid phase of a crystal slurry without interference of the dispersed crystals.

Fourier transform (FT) spectrometers have a number of advantages over dispersive instruments, that is, reduced measuring time and increased light throughput, hence a better signal-to-noise ratio. Basically, a FT spectrometer is a Michelson interferometer where the spectrum is reconstructed using a FT of the interference pattern of the measured sample (Schrader, 1995). A FT instrument allows us to measure all wavelengths at once while in a dispersive instrument, a monochromatic beam changes its wavelength over time. Thus, the overall measuring time is shorter in a FT spectrometer as compared to a dispersive instrument.

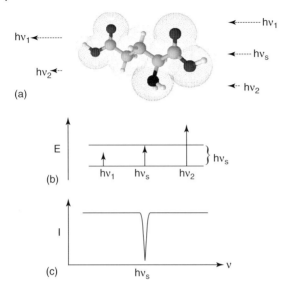

Figure 9.1 Principle of IR absorption: (a) Quanta of energy $h\nu$ impacts the molecule (L-glutamic acid) resulting in elastic scattering or absorption; (b) energy level diagram: photon of frequency of ν_S is absorbed; and (c) simplified IR absorption spectrum (Schrader, 1995).

9.2 Calibration

The law of Lambert–Beer is commonly used in infrared spectroscopy to relate the measured absorbance A to the concentration c of the sample:

$$A = \varepsilon l c \tag{9.1}$$

where ε is the extinction coefficient and l is the path that the light travels through the material (Schrader, 1995). This law is usually valid only at low concentrations and deviations at high concentrations might be attributed to effects from the sample itself, for example, deviations arising from intermolecular interactions. At low concentrations, the measurement error is large due to low signal intensity and low signal-to-noise ratio. The optimum range for quantitative analysis of IR spectra is in the range of $0.2 < A < 0.7$ (Schrader, 1995).

The peak height or area of a characteristic peak in the IR spectrum can be related to the concentration of the solute. However, a univariate approach is less accurate and robust than a multivariate data analysis where the whole spectrum is included. The manual selection of characteristic peaks can be complex and time consuming (Togkalidou et al., 2002). Additionally, characteristic bands of different molecules can be overlapping and peak deconvolution techniques are required. A major problem in measuring concentrations in crystallization processes arises from difficulties in de-coupling absorption bands of the solute from the background of the solvent. This is usually overcome by obtaining the difference spectra between the solute and the solvent. In nonisothermal crystallization processes, both the solute concentration and the temperature are changing. This makes the application of a univariate approach even more difficult since the absorbance spectra depend not only on concentration but also on temperature. Moreover, it is recommended to subtract the baseline from the original data to avoid the effect of baseline drifts, which are often encountered.

Multivariate methods automatically assign weight factors to all wavenumbers, thus making peak selection and peak deconvolution techniques redundant. In spectroscopic studies, many wavenumbers are correlated and projection-based methods such as principal component regression (PCR) and partial least-squares regression (PLSR) are applied to solve the ill-conditioned regression problem (Cornel, Lindenberg, and Mazzotti, 2008; Geladi and Kowalski, 1986; Nadler and Coifman, 2005). In Figure 9.2, a typical calibration procedure is shown (Cornel, Lindenberg, and Mazzotti, 2008). An undersaturated solution with known concentration is cooled until nucleation is detected, either visually or by means of a turbidity or focused beam reflectance measurement (FBRM) probe, and the spectra are recorded. This procedure is repeated for different initial concentrations as illustrated in Figure 9.2. The recorded data form the calibration set, which is employed to develop a calibration model using PCR or PLSR (Cornel, Lindenberg, and Mazzotti, 2008).

The accuracy of the concentration measurement depends very much on the properties of the compound and of the solvent, on the calibration model applied, and on the devices used. Changes of the medium composition in the optical conduit,

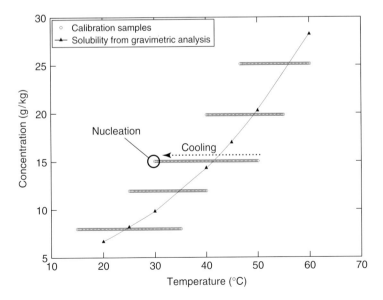

Figure 9.2 Calibration procedure.

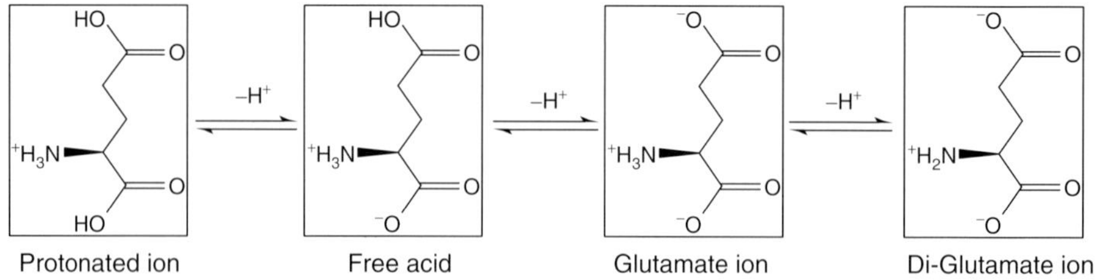

Figure 9.3 Species of L-glutamic acid in aqueous solution.

for example, water vapor or CO_2, are known to have effects on the sensitivity. Also, small deviations of the solvent composition from the reference composition used for calibration, for example, varying amounts of water, might lead to erroneous calculations of the concentration.

9.3
Speciation Monitoring

Many organic substances exist as several co-existing species in solution, for example, L-glutamic acid, which reacts in aqueous solutions to yield four different species, as shown in Figure 9.3 (Schöll et al., 2006b). The species concentrations depend on the pH of the solution: the di-glutamate ion prevails at high pH values, while reducing the pH gives rise to the glutamate ion. The free acid and the protonated ions are formed by protonation reactions.

Figure 9.4a shows the recorded IR spectra for different pH values and highlights the characteristic bands corresponding to the functional groups of L-glutamic acid. It can be seen that different bands appear at different pH values. The identification of these bands allows for a determination of the species concentration based on the absorbance data and by applying the law of Lambert–Beer (Schöll et al., 2006b).

The measured equivalent concentrations of the functional groups are shown in Figure 9.4b. The results from a speciation model of L-glutamic acid, which is based on the dissociation equilibria, are shown as well. It is worth noting that there is a good agreement with the experimental data.

9.4
Co-Crystal Formation

The formation of co-crystals of two or more compounds in a fixed stoichiometric ratio is a new pathway for the crystallization of otherwise difficult to crystallize substances. Furthermore, co-crystals may have enhanced physicochemical properties in comparison to crystals of the original substance. ATR-FTIR spectroscopy

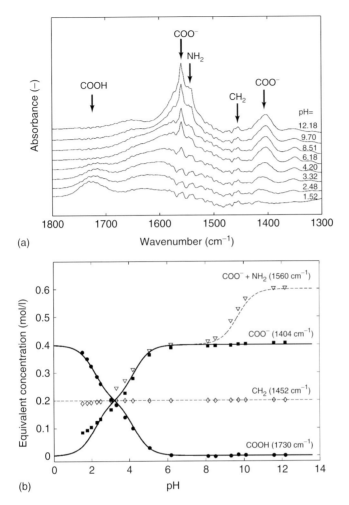

Figure 9.4 IR spectra of L-glutamic acid in aqueous solutions at different pH values (a) Experimental (symbols) and simulated (lines) equivalent concentration of different functional groups as a function of pH. (b) The nominal concentration of L-glutamic acid was 0.2 mol/l. The equivalent concentration of CH_2 is divided by 2 for better readability (Schöll et al., 2006b).

can be used to identify the concentrations of the two co-crystal forming substances independently. In the case of carbamazepine and nicotinamide, it was found that the formation of the co-crystals of carbamazepine/nicotinamide (1 : 1) is in competition with the formation of pure carbamazepine crystals (Gagniere et al., 2009). Since the stoichiometric ratio of the co-crystal is known, the formation of co-crystals should change the concentrations of the two compounds in that ratio. The crystallization of pure carbamazepine crystals however only depletes the concentration of carbamazepine in solution. Measuring the concentration changes of the two drug substances, it was hence possible to determine how much of the two crystalline phases was formed.

Furthermore, the use of ATR-FTIR spectroscopy allows determining complete temperature-dependent phase diagrams for solute/solute/solvent systems, as was shown for the co-crystallization of caffeine and glutaric acid from acetonitrile (Yu, Chow, and Tan, 2010). The knowledge of such a complete phase diagram allows defining safe operating regions where the co-crystal is consistently produced with high purity.

9.5
Solubility Measurement

ATR-FTIR spectroscopy can be used to measure solubilities, for example, as a function of temperature or solvent composition. In the case of the β-polymorph of L-glutamic acid, a saturated suspension with an excess of L-glutamic acid crystals was heated slowly from 20 to 60 °C, as shown in Figure 9.5a (Cornel, Lindenberg, and Mazzotti, 2008). Temperature plateaus were applied to verify equilibrium conditions. The measured solubility curve is shown in Figure 9.5b together with gravimetrically determined solubilities. A very good agreement can be observed despite the low concentrations. This approach enables nonisothermal measurement of the solubility and was applied successfully to different other substances. Due to its measurement principle, concentrations as low as 1 wt% can be measured with a relatively high accuracy of ±1% (Cornel, Lindenberg, and Mazzotti, 2008).

9.6
Crystal Growth Rates

The change of supersaturation over time, that is, the desupersaturation curve, can be monitored using ATR-FTIR. For specifically designed seeded isothermal desupersaturation experiments, the IR data can be used in conjunction with additional experimental data, such as seed mass and initial particle size distribution, to calculate crystal growth rates. The desupersaturation curves of a seeded batch experiment of α L-glutamic acid in aqueous solution are shown in Figure 9.6 (Schöll et al., 2007b). Three experiments with different seed sizes were conducted. The other operating conditions such as initial concentration and seed mass were kept constant. An increase of the total surface area of the seed crystals by reducing seed size leads to a faster decrease in supersaturation. On the basis of such experiments, the growth kinetics can be estimated using a population balance model and an optimization procedure which minimizes the error between experimental and simulated supersaturation values (Schöll et al., 2007b).

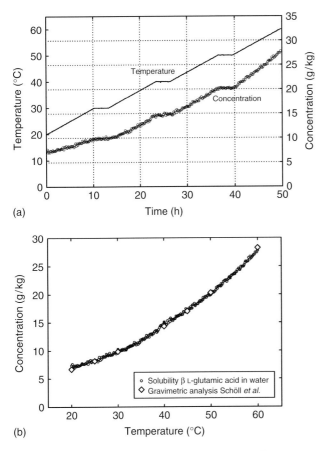

Figure 9.5 Measured concentration of β L-glutamic acid as a function of temperature: (a) solubility of β L-glutamic acid as a function of temperature. (b) (Cornel, Lindenberg, and Mazzotti, 2008). The concentration units are grams of solute per kilogram of solvent.

9.7
Polymorph Transformation

Polymorphism, that is, the ability of a substance to crystallize in different crystal structures, is of high importance in the pharmaceutical industry since it results in different physical and chemical properties such as stability, solubility, and reactivity (Cardew, Davey, and Ruddick, 1984). The solvent-mediated polymorph transformation of the metastable α to the stable β polymorph of L-glutamic acid was monitored using ATR-FTIR as shown in Figure 9.7 (Schöll et al., 2006a).

In the first phase, the α form nucleates from the supersaturated solution. The crystals grow until the supersaturation is consumed; the concentration reaches the solubility limit of the α polymorph (Phase 2). Subsequently, the β polymorph

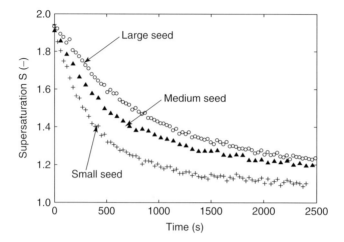

Figure 9.6 Desupersaturation curves of α L-glutamic acid for different seed sizes (Schöll et al., 2007b).

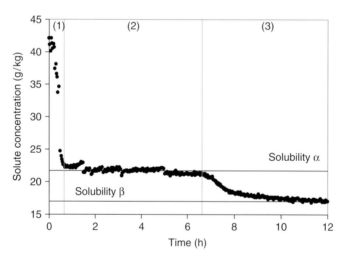

Figure 9.7 Measured concentration of L-glutamic acid during polymorph transformation (Schöll et al., 2006a). The concentration units are grams of solute per kilogram of solvent.

nucleates and grows, whereas the α polymorph dissolves. In this phase, the concentration stays constant since the dissolution of α is much faster than the growth of β. It must be noted that the scattering of the concentration in this phase stems from the presence of solid particles on the probe window. After the last α crystal has dissolved, the concentration decreases to the solubility of the β polymorph (Phase 3).

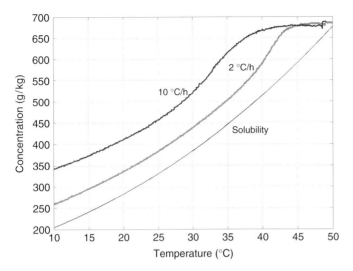

Figure 9.8 Cooling crystallization of ascorbic acid at different cooling rates. The concentration units are grams of solute per kilogram of solvent.

9.8
Crystallization Monitoring and Control

ATR-FTIR can be applied to monitor the supersaturation, which is a prerequisite for model development, design, and control of crystallization processes (Fujiwara et al., 2002; Groen and Roberts, 2001; Lewiner et al., 2001; Zhou et al., 2006). The effect of different cooling rates on the course of the supersaturation is shown in Figure 9.8 for the cooling crystallization of ascorbic acid in aqueous solutions (Eggers, Kempkes, and Mazzotti, 2008). Saturated solutions were cooled down at different linear cooling rates. It can be seen that for higher cooling rates higher supersaturations can be achieved.

9.9
Impurity Monitoring

ATR-FTIR can be used to monitor the level of impurities and formation of undesired side-products during crystallization processes (Derdour et al., 2003; Lin et al., 2006). Impurities can have tremendous effects on the systems thermodynamics and kinetics. Therefore, impurity monitoring is of particular interest to the pharmaceutical industry, where small concentrations of similar molecules are often present during the crystallization step of new active pharmaceutical ingredients. In Derdour et al. (2003), impurity concentrations as low as 300 ppm could be measured using ATR-FTIR, though it must be noted that the signal-to-noise ratio was relatively low.

9.10
Conclusions

ATR-FTIR spectroscopy can be employed to measure liquid phase concentrations, for example, for *in situ* solubility determination, growth rate estimation, polymorph transformation, speciation, co-crystal formation, and impurity monitoring. Better calibration models are obtained by multivariate analysis of ATR-FTIR data as compared to univariate approaches, for example, based solely on the peak height or area. ATR-FTIR spectroscopy in combination with multivariate data analysis is of great importance for an accurate *in situ* measurement of the liquid-phase composition, especially when measuring in a suspension as in crystallization processes.

References

Cardew, P.T., Davey, R.J., and Ruddick, A.J. (1984) Kinetics of polymorphic solid-state transformations. *J. Chem. Soc. – Faraday Trans.*, **80**, 659–668.

Cornel, J., Lindenberg, C., and Mazzotti, M. (2008) Quantitative application of *in-situ* ATR-FTIR and Raman spectroscopy in crystallization processes. *Ind. Eng. Chem. Res.*, **47** (14), 4870–4882.

Derdour, L., Fevotte, G., Puel, F., and Carvin, P. (2003) Real-time evaluation of the concentration of impurities during organic solution crystallization. *Powder Technol.*, **129** (1–3), 1–7.

Eggers, J., Kempkes, M., and Mazzotti, M. (2008) Monitoring size and shape during cooling crystallization of ascorbic acid. *Chem. Eng. Sci.*, **64** (1), 163–171.

Fujiwara, M., Chow, P.S., Ma, D.L., and Braatz, R.D. (2002) Paracetamol crystallization using laser backscattering and ATR-FTIR spectroscopy: metastability, agglomeration, and control. *Cryst. Growth Des.*, **2** (5), 363–370.

Gagniere, E., Mangin, D., Puel, F., Bebon, C., Klein, J.-P., Monnier, O., and Garcia, Eric. (2009) Cocrystal formation in solution: *in situ* solute concentration monitoring of the two components and kinetic pathways. *Cryst. Growth Des.*, **9** (8), 3376–3383.

Geladi, P. and Kowalski, B.R. (1986) Partial least-squares regression – a tutorial. *Anal. Chim. Acta*, **185**, 1–17.

Groen, H. and Roberts, K.J. (2001) Nucleation, growth, and pseudo-polymorphic behavior of citric acid as monitored *in situ* by attenuated total reflection Fourier transform infrared spectroscopy. *J. Phys. Chem. B*, **105** (43), 10723–10730.

Lewiner, F., Klein, J.P., Puel, F., and Fevotte, G. (2001) On-line ATR FTIR measurement of supersaturation during solution crystallization processes. Calibration and applications on three solute/solvent systems. *Chem. Eng. Sci.*, **56** (6), 2069–2084.

Lin, Z.H., Zhou, L.L., Mahajan, A., Song, S., Wang, T., Ge, Z.H., and Ellison, D. (2006) Real-time endpoint monitoring and determination for a pharmaceutical salt formation process with in-line FT-IR spectroscopy. *J. Pharm. Biomed. Anal.*, **41** (1), 99–104.

Nadler, B. and Coifman, R.R. (2005) Partial least squares, Beer's law and the net analyte signal: statistical modeling and analysis. *J. Chemom.*, **19** (1), 45–54.

Schöll, J., Bonalumi, D., Vicum, L., Mazzotti, M., and Müller, M. (2006a) In situ monitoring and modeling of the solvent-mediated polymorphic transformation of L-glutamic acid. *Cryst. Growth Des.*, **6** (4), 881–891.

Schöll, J., Vicum, L., Müller, M., and Mazzotti, M. (2006b) Precipitation of L-glutamic acid: determination of nucleation kinetics. *Chem. Eng. Technol.*, **29** (2), 257–264.

Schöll, J., Lindenberg, C., Vicum, L., Brozio, J., and Mazzotti, M. (2007a) Antisolvent precipitation of PDI 747: kinetics of particle formation and growth. *Cryst. Growth Des.*, **7** (9), 1653–1661.

Schöll, J., Lindenberg, C., Vicum, L., Brozio, J., and Mazzotti, M. (2007b) Precipitation of L-glutamic acid: determination of growth kinetics. *Faraday Discuss.*, **136**, 238–255.

Schrader, B. (1995) *Infrared and Raman Spectroscopy: Methods and Applications*, Wiley-VCH Verlag GmbH, Weinheim.

Shriver, D.F. and Atkins, P.W. (1999) *Inorganic Chemistry*, Oxford University Press, Oxford.

Togkalidou, T., Tung, H.H., Sun, Y.K., Andrews, A., and Braatz, R.D. (2002) Solution concentration prediction for pharmaceutical crystallization processes using robust chemometrics and ATR FTIR spectroscopy. *Org. Process Res. Dev.*, **6** (3), 317–322.

Yu, Z.Q., Chow, P.S., and Tan, R.B.H. (2010) Operating regions in cooling cocrystallization of caffeine and glutaric acid in acetonitrile. *Cryst. Growth Des.*, **10**, 2382–2387.

Zhou, G.X., Fujiwara, M., Woo, X.Y., Rusli, E., Tung, H.H., Starbuck, C., Davidson, O., Ge, Z.H., and Braatz, R.D. (2006) Direct design of pharmaceutical antisolvent crystallization through concentration control. *Cryst. Growth Des*, **6** (4), 892–898.

10
Raman Spectroscopy

Jeroen Cornel, Christian Lindenberg, Jochen Schöll, and Marco Mazzotti

10.1
Introduction

Contrary to infrared spectroscopy, monochromatic light is used to irradiate a sample in Raman spectroscopy. The Raman effect is due to the inelastic scattering of incident light. If a light quantum or a photon with energy $h\nu_0$ hits a molecule, light can be scattered elastically, that is, the scattered photon has the same energy, or inelastically, that is, the energy carried by the scattered photon has changed with respect to the incoming photon. The elastic scattering process has the highest probability and is known as *Rayleigh scattering*. However, at a lower probability also the inelastic, so-called Raman scattering process, occurs and the resulting scattered energy quantum has an energy of $h\nu_0 \pm h\nu_s$, where $h\nu_s$ is related to the molecular structure of the compound. The Raman scattered light is frequency shifted with respect to the excitation frequency to lower or to higher frequencies resulting in Stokes or anti-Stokes Raman scattering, respectively. The principle of Raman scattering is illustrated in Figure 10.1. A detailed description may be found in Long (1977).

At ambient temperature, most molecules are in their vibrational ground state. According to Boltzmann's law, a much smaller number of molecules are in the vibrational excited state. Therefore, Raman scattering resulting in a quantum with lower energy $h\nu_0 - h\nu_s$ has a higher probability than the reverse process, that is, emission of a quantum with higher energy corresponding to $h\nu_0 + h\nu_s$. Therefore, the Stokes signal has a higher intensity than the anti-Stokes signal as illustrated in Figure 10.1 (Schrader, 1995) (at ambient temperatures). The Raman scattering effect is so feeble that only about one photon in every 10^{12} incident photons is scattered inelastically. However, the use of intense laser radiation and very efficient photomultiplier detectors make this technique viable. Typically, the lasers used in Raman spectrometers emit light in the near-infrared range. Mostly, Raman spectrometers are equipped with conventional CCD detectors or employ Fourier transform to record the data. Raman spectroscopy can be applied off-line as well as *in situ* to monitor a wide variety of chemical processes.

Industrial Crystallization Process Monitoring and Control, First Edition. Edited by
Angelo Chianese and Herman J. M. Kramer.
© 2012 Wiley-VCH Verlag GmbH & Co. KGaA. Published 2012 by Wiley-VCH Verlag GmbH & Co. KGaA.

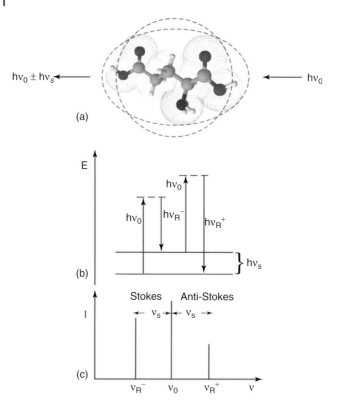

Figure 10.1 Principle of Raman scattering. (a) Quanta of energy $h\nu_0$ hit the molecule (L-glutamic acid) resulting in inelastic scattering; (b) energy level diagram: irradiation with light quanta $h\nu_0$ may result in scattering of quanta with energy $h\nu_R^- = h\nu_0 - h\nu_s$ or $h\nu_R^+ = h\nu_0 + h\nu_s$, Stokes and anti-Stokes scattering, respectively; (c) simplified Raman spectrum, signal at ν_0 is due to Rayleigh scattering, signal at lower frequency (Stokes signal) has a higher intensity than the signal at higher frequency (anti-Stokes signal) (Schrader, 1995).

10.2
Factors Influencing the Raman Spectrum

Raman scattering results from both the solid and the liquid phase as well as from interactions between those phases; hence, properties of both phases have to be considered for the application of Raman spectroscopy in heterogeneous processes such as crystallization and precipitation.

Concerning liquid-phase properties, the solvent composition has a large impact on the Raman spectrum. Depending on the wavelength of the laser source, water can exhibit relatively low Raman scattering; therefore, it can be an ideal solvent for Raman spectroscopy studies. However, numerous solvents exhibit strong Raman scattering and subtracting a solvent background is not always possible, which can make the application of Raman spectroscopy challenging.

Several solid-phase properties are known to influence the Raman scattering effect, among them are solid-phase composition, particle size, and shape. Whereas the solid-phase composition results in different signals at different characteristic Raman shifts, the particle properties scale the intensity of different characteristic signals. For this reason, Raman spectroscopy is mostly employed to extract information about the composition of the solid phase, that is, in the case of (pseudo-)polymorphism. Particle size and shape effects on Raman spectroscopy have hardly been studied and it has been proposed that the particle size effects depend on the type of optics being used (Strachan *et al.*, 2007; Hu *et al.*, 2006; Wang, Mann, and Vickers, 2002).

The suspension density has also a pronounced effect on Raman signal intensity; for relatively low suspension densities, the intensity is increasing linearly with the solid concentration. Temperature influences the Raman signal intensity as well; however, for processes at ambient conditions, its effect can typically be neglected (Schrader, 1995).

10.3
Calibration

The presence of a signal at a particular wavenumber or Raman shift is due to the presence of a particular functional group in a molecule or a cluster of molecules. In principle, the molecular structure can be deduced from a Raman spectrum; however the complex interaction between liquid and solid properties makes this application of Raman spectroscopy particularly challenging and is therefore rarely reported (Shriver and Atkins, 2001; Atkins, 1999; Shurvell and Bergin, 1989).

10.3.1
Univariate Approaches

Quantitative application of any kind of measurement technique requires a model to relate the measured (dependent) variables, for example, Raman intensity at different Raman shifts, to the independent variables, for example, concentration or solid-phase composition. In spectroscopy, a relationship between peak height or peak area and concentration can be applied. By selecting different characteristic signals and measuring samples of known compositions, a quantitative model, the calibration, can be developed. However, the effect of different process conditions on Raman signal intensity makes this method challenging. An interesting approach is the quantification of the solid composition based on the position of a characteristic signal (Wang *et al.*, 2000). This method can be employed if the two signals, corresponding to two different solid forms, are separated by a minor Raman shift. However, the effect of different process parameters and the possibility of overlapping signals make this approach not always feasible. Figure 10.2a displays the characteristic peaks for the α and β polymorph and for the solute in the case

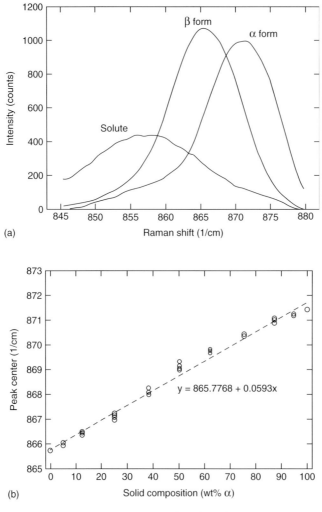

Figure 10.2 (a) Peaks characterizing the solute, the α and β polymorph and (b) calibration based on measurements using dry powder mixtures of α and β forms.

of L-glutamic acid. As can be seen, the α and β signals are separated only by 5–6 cm^{-1} and the position of the resulting peak, that is, the sum of the three signal as shown in Figure 10.2a, can be used to quantify the solid composition as shown in Figure 10.2b.

The course of the solid composition over time, estimated based on peak position, is shown for an unseeded polymorph transformation experiment of L-glutamic acid in Figure 10.3a. It can be observed that this approach gives reasonable results for relatively high initial supersaturation resulting in high suspension densities.

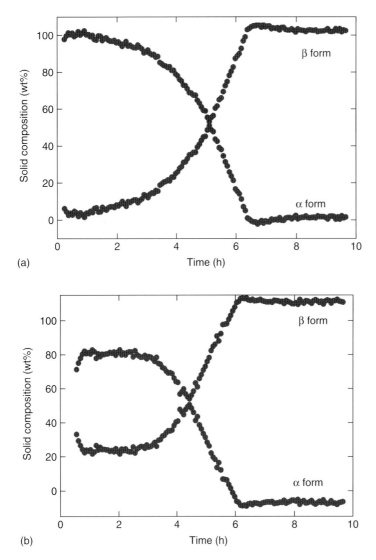

Figure 10.3 Solid composition profiles for unseeded polymorph transformations at (a) high ($S_\alpha = 3.5$) and (b) low initial supersaturation ($S_\alpha = 1.75$).

Thus, the intensity of the solute signal is low as compared to the signal intensity due to the solid phase and the effect on the position of the signal is negligible. For lower initial supersaturations resulting in lower suspension densities, the solute signal shifts the position of the peak toward lower Raman shifts resulting in an underestimation of the solid composition in terms of weight fraction of the metastable α polymorph, as shown in Figure 10.3b.

10.3.2
Multivariate Approaches

Univariate approaches do not always meet the accuracy and robustness that can be achieved by utilizing a broader range of the measured spectra through the application of multivariate approaches. Moreover, the manual selection of characteristic signals can be complex and time consuming. Besides, signals can be overlapping and complex peak deconvolution techniques are required. Multivariate methods automatically assign weight factors to all included variables thus making variable selection and peak deconvolution techniques redundant. As in infrared spectroscopy, the Raman signal intensity scales linearly with the amount of scattering material per unit volume for relatively low suspension densities. This scattering material can be present both in the solution and in the solid phase in the case of suspensions (Schrader, 1995). Therefore, linear models can be used to describe the relationship between the signal intensity and the concentration for both infrared and Raman spectroscopy. The objective is to regress the concentrations on the spectral data (calibration) and to use the regression vector to estimate the unknown concentrations in a series of new measurements. Figure 10.4 shows typical Raman spectra for suspensions of L-glutamic acid with different solid-phase compositions. Multivariate data analysis employs a large part of the measured spectra to estimate the solid- and liquid-phase composition (Cornel and Mazzotti, 2008, 2009).

Among numerous multivariate data analysis techniques, principal component regression (PCR) and partial least-squares regression (PLSR) are applied most

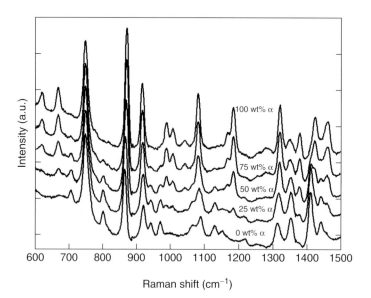

Figure 10.4 Raman spectra for suspensions of constant suspension density at 25 °C with different solid compositions.

frequently. The discussion about these techniques is beyond the scope of this work and can be found elsewhere (Smilde, Bro, and Geladi, 2004).

10.4 Applications

10.4.1 Solid-Phase Composition Monitoring

The main application of Raman spectroscopy in crystallization processes is found in the quantification of the solid-phase composition. Raman spectroscopy can be employed off-line as well as *in situ* monitoring of the solid-phase composition of various processes, which is of increasing importance in the fine-chemical and pharmaceutical industry. Most of the scientific studies discuss the characterization of solids in the case of dry powder or tablets; see Strachan *et al.* (2007) for an excellent review.

One of the first publications that discussed *in situ* polymorph transformation monitoring in suspensions by means of Raman spectroscopy employs the position of the characteristic signal to quantify the solid-phase composition (Wang *et al.*, 2000). After that, more groups applied Raman spectroscopy to monitor solid-state transformation processes through the application of different calibration methods (Ono, ter Horst, and Jansens, 2004; Schöll *et al.*, 2006; Qu *et al.*, 2006; Caillet, Puel, and Fevotte, 2006; Caillet *et al.*, 2007; Hu *et al.*, 2005; Starbuck *et al.*, 2002; Falcon and Berglund, 2004). Besides solid-state characterization at ambient conditions, specially equipped Raman probes can be used to monitor *in situ* crystallization and precipitation processes at elevated pressures (Wolf *et al.*, 2004; Hänchen *et al.*, 2008). Figure 10.5 displays the time-resolved Raman spectra for the transformation of hydromagnesite to magnesite at 120 °C and 100 bar (Hänchen *et al.*, 2008).

10.4.2 Liquid Phase Composition Monitoring

The Raman scattering effect is caused by the solid as well as the liquid phase; therefore, it is in principle possible to extract liquid-phase information from the Raman spectra, as shown in the literature (Caillet, Puel, and Fevotte, 2006; Caillet *et al.*, 2007; Hu *et al.*, 2005; Tamagawa, Miranda, and Berglund, 2002). Different methods have been applied: one approach is to relate the intensity of a certain signal to the absolute suspension density and calculate the solute concentration through the mass balance of the process (Caillet, Puel, and Fevotte, 2006; Caillet *et al.*, 2007). Another approach is to select the signal attributed to the solute and use its height or area to estimate the solute concentration directly (Hu *et al.*, 2005; Tamagawa, Miranda, and Berglund, 2002). Recently, it was shown that Raman spectroscopy combined with multivariate data analysis enables solute concentration estimation despite the fact that the solute signals are weak and

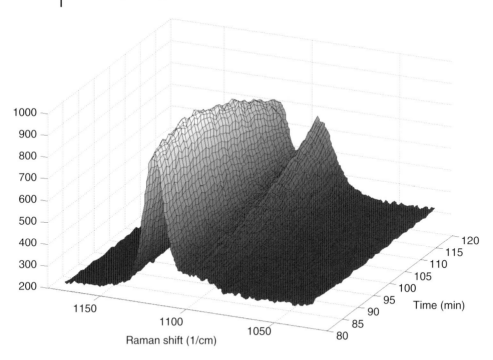

Figure 10.5 Time-resolved Raman spectra of transformation of hydromagnesite to magnesite at 120 °C and 100 bar.

completely overlapping with signals related to the solid phase (Cornel, Lindenberg, and Mazzotti, 2008). Figure 10.6 shows a representative unseeded polymorph transformation experiment during which the metastable α polymorph of L-glutamic acid precipitates and transforms into the stable β polymorph (Cornel, Lindenberg, and Mazzotti, 2008). The time-resolved spectra clearly show the different phases of the transformation. The formation of crystals can be readily observed by the clear change in the Raman spectra during the first hour of the experiment, that is, a strong increase in Raman intensity at 870 cm^{-1} due to the solid phase. Moreover, the depletion of supersaturation can be observed by a strong decrease in Raman intensity in the characteristic range for the solute, that is, at 857 cm^{-1}. Multivariate data analysis enables solid- and liquid-phase composition monitoring; its outcome can be observed in Figure 10.6b.

10.4.3
Amorphous Content Quantification

Several studies have suggested the application of Raman spectroscopy to quantify the crystallinity of pharmaceutical solids. Assuming that 100% crystalline material is available, univariate as well as multivariate methods have shown that Raman spectroscopy enables the determination of the crystallinity as low as several weight percents (Strachan et al., 2007).

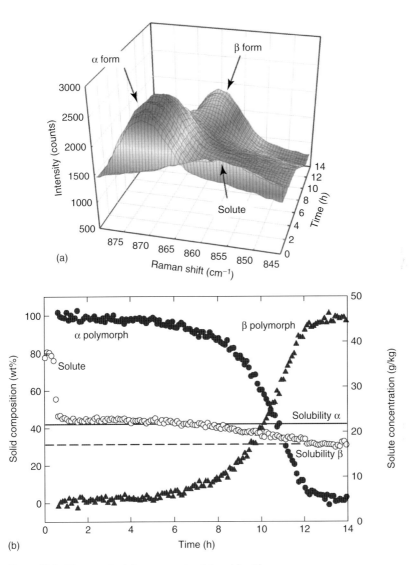

Figure 10.6 Time-resolved Raman spectra (a) and liquid- and solid-phase compositions (b) for an unseeded solid-state transformation of L-glutamic acid.

10.5 Conclusions

The Raman scattering signal emerges from both the liquid and solid phases (of course, also molecules in the gas phase may show Raman activity but due to the low density the signal is rather weak); hence, there are numerous factors influencing the signal, which makes the quantitative application of this technique more challenging. Using more advanced data analysis techniques, Raman spectroscopy can be employed to estimate the liquid- as well as the solid-phase composition in heterogeneous processes such as crystallization and precipitation.

References

Atkins, P.W. (1999) *Physical Chemistry*, 6th edn, Oxford University Press, Oxford.

Caillet, A., Puel, F., and Fevotte, G. (2006) In-line monitoring of partial and overall solid concentration during solvent-mediated phase transition using Raman spectroscopy. *Int. J. Pharm.*, **307** (2), 201–208.

Caillet, A., Rivoire, A., Galvan, J.M., Puel, F., and Fevotte, G. (2007) Crystallization of monohydrate citric acid. 1. In situ monitoring through the joint use of Raman spectroscopy and image analysis. *Cryst. Growth Des.*, **7** (10), 2080–2087.

Cornel, J., Lindenberg, C., and Mazzotti, M. (2008) Quantitative Application of in Situ ATR-FTIR and Raman Spectroscopy in Crystallization Processes, *Ind. Eng. Chem. Res.*, **47** (14), 4870–4882.

Cornel, J. and Mazzotti, M. (2008) Calibration-free quantitative application of *in situ* Raman spectroscopy to a crystallization process. *Anal. Chem.*, **80**, 9240–9249.

Cornel, J. and Mazzotti, M. (2009) Estimating crystal growth rates using *in situ* ATR-FTIR and Raman spectroscopy in a calibration-free manner. *Ind. Eng. Chem. Res.*, **48**, 10740–10745.

Falcon, J.A. and Berglund, K.A. (2004) In situ monitoring of antisolvent addition crystallization with principal components analysis of Raman spectra. *Cryst. Growth Des.*, **4** (3), 457–463.

Hänchen, M., Prigiobbe, V., Baciocchi, R., and Mazzotti, M. (2008) Precipitation in the Mg–carbonate system – effects of temperature and CO_2 pressure. *Chem. Eng. Sci.*, **63** (4), 1012–1028.

Hu, Y.R., Liang, J.K., Myerson, A.S., and Taylor, L.S. (2005) Crystallization monitoring by Raman spectroscopy: simultaneous measurement of desupersaturation profile and polymorphic form in flufenamic acid systems. *Ind. Eng. Chem. Res.*, **44** (5), 1233–1240.

Hu, Y.R., Wikstrom, H., Byrn, S.R., and Taylor, L.S. (2006) Analysis of the effect of particle size on polymorphic quantification by Raman spectroscopy. *Appl. Spectrosc.*, **60** (9), 977–984.

Long, D.A. (1977) *Raman Spectroscopy*, McGraw-Hill, London.

Ono, T., ter Horst, J.H., and Jansens, P.J. (2004) Quantitative measurement of the polymorphic transformation of L-glutamic acid using *in situ* Raman spectroscopy. *Cryst. Growth Des.*, **4** (3), 465–469.

Qu, H.Y., Louhi-Kultanen, M., Rantanen, J., and Kallas, J. (2006) Solvent-mediated phase transformation kinetics of an anhydrate/hydrate system. *Cryst. Growth Des.*, **6** (9), 2053–2060.

Schöll, J., Bonalumi, D., Vicum, L., Mazzotti, M., and Muller, M. (2006) *In situ* monitoring and modeling of the solvent-mediated polymorphic transformation of L-glutamic acid. *Cryst. Growth Des.*, **6** (4), 881–891.

Schrader, B. (1995) *Infrared and Raman Spectroscopy: Methods and Applications*, Wiley-VCH Verlag GmbH, Weinheim.

Shriver, D.F. and Atkins, P.W. (2001) *Inorganic Chemistry*, 3rd edn, Oxford University Press, Oxford.

Shurvell, H.F. and Bergin, F.J. (1989) Raman-spectra of L(+)-glutamic acid and related-compounds. *J. Raman Spectrosc.*, **20** (3), 163–168.

Smilde, A.K., Bro, R., and Geladi, P. (2004) *Multi-way Analysis with Application in Chemical Sciences*, John Wiley & Sons, Ltd, Chichester, UK.

Starbuck, C., Spartalis, A., Wai, L., Wang, J., Fernandez, P., Lindemann, C.M., Zhou, G.X., and Ge, Z.H. (2002) Process optimization of a complex pharmaceutical polymorphic system via *in situ* Raman spectroscopy. *Cryst. Growth Des.*, **2** (6), 515–522.

Strachan, C.J., Rades, T., Gordon, K.C., and Rantanen, J. (2007) Raman spectroscopy for quantitative analysis of pharmaceutical solids. *J. Pharm. Pharmacol.*, **59** (2), 179–192.

Tamagawa, R.E., Miranda, E.A., and Berglund, K.A. (2002) Raman spectroscopic monitoring and control of aprotinin supersaturation in hanging-drop crystallization. *Cryst. Growth Des.*, **2** (4), 263–267.

Wang, H.L., Mann, C.K., and Vickers, T.J. (2002) Effect of powder properties on the intensity of Raman scattering by crystalline solids. *Appl. Spectrosc.*, **56** (12), 1538–1544.

Wang, F., Wachter, J.A., Antosz, F.J., and Berglund, K.A. (2000) An investigation of solvent mediated polymorphic transformation of progesterone using *in situ* Raman spectroscopy. *Org. Process Res. Dev.*, **4** (5), 391–395.

Wolf, G.H., Chizmeshya, A.V.G., Diefenbacher, J., and McKelvy, M.J. (2004) *In situ* observation of CO_2 sequestration reactions using a novel microreaction system. *Environ. Sci. Technol.*, **38** (3), 932–936.

11
Basic Recipe Control
Alex N. Kalbasenka, Adrie E.M. Huesman, and Herman J.M. Kramer

11.1
Introduction

In basic recipe control, expert knowledge and practical experience with the considered batch crystallization process are combined with available experimental data to form a set of heuristic rules. Those rules are used as guidelines for defining a recipe and implementing it in a basic control algorithm. Laboratory-scale experiments are often used to quantify the properties of the system and to study the process behavior. The experimental data is used to scale up the process. No process modeling is involved in defining the basic recipe control strategy. Some calculations using a process model might be used to either define a starting point for the recipe design or guide the experimental search in optimization of the recipe.

11.2
Incentives for Basic Recipe Control

In general, the goal of crystallization control is twofold. On the one hand, it is desired to produce the product of a desired quality and quantity in a consistent and reproducible way. On the other hand, crystallization step should facilitate downstream processes during which ideally the product properties should be preserved (Kim *et al.*, 2005). The interaction between the crystallization step and the following product processing steps is crucial for the overall performance of a production process. Depending on a particular application, the dominant control objective is dictated either by the product characteristics due to imposed product specifications (optimization of the crystallization step, often happens when the crystallized material is an end product) or by the operational reasons (crystallization control for optimization of the downstream processing steps, an intermediate product is produced).

As far as the crystallization step is concerned, the control objective is often formulated in terms of the characteristic crystal size, the crystal size distribution (CSD), the crystal purity, the process yield, or the batch time. Other product properties such as the dissolution rate, the crystal habit, and ability of the crystalline material to flow are functions of the above-mentioned characteristics. If these control objectives are not met during crystallization step of the process, downstream operations must include additional processing steps such as milling, sieving, or even (partial) recrystallization of the product.

In the case, when product characteristics do not play the dominant role in defining the control objectives (often happens if the crystallized material is an intermediate product), the entire production process is optimized to remove bottlenecks. The limiting steps are often among the downstream processes such as filtration, washing, and drying. A limited yield or a long batch time can make crystallization a limiting step too. As for the operational incentives for controlling the crystallization process, ease and efficiency of the downstream and handling processes are often assessed in terms of characteristic parameters of each subsequent processing/handling step: solid–liquid separation, washing, and drying (cake resistance, cake porosity, and filtration time (Togkalidou et al., 2001)), resizing (milling or granulation), transportation (flowability), storage behavior (tendency to caking), and usage (dusting). Large crystals with a narrow CSD are usually required to improve filterability, crystal purity, and storage behavior of the crystalline product.

In practice, a combination of the objectives is required as is often desired to achieve both an acceptable efficiency of the production process (processing time and product yield) and the required product characteristics. Depending on the product, requirements on either the process yield or the crystal properties can prevail.

11.3
Main Mechanisms, Sensors, and Actuators

11.3.1
Crystallization Mechanisms

The supersaturation is the driving force of crystallization processes. Rates of nucleation, growth, and agglomeration and as a consequence the crystal morphology, the CSD, the crystal purity are complex functions of the supersaturation. Controlling the product properties implies manipulating the supersaturation to achieve a favorable balance among nucleation, growth, and agglomeration rates. If the balance is not in favor of crystal growth, then controlling a batch means dealing with the consequences of uncontrolled particle formation processes.

11.3.2
Sensors

Measurement of the critical process variables is essential for the process analysis, monitoring, and control. Recent developments of process analytical technology

(PAT) tools offer new perspectives for understanding, optimizing, and controlling crystallization processes. The PAT initiative was introduced in various sectors of food and pharmaceutical industry to facilitate the shift from regulating the process to regulating the product for insuring sufficient quality and consistency of the product (Yu *et al.*, 2007). The main purpose of the PAT initiative is to improve the process development and control through the process analysis by using real-time (online and in-line) analyzers (Bakeev, 2005). There is a growing interest for application of the PAT tools for real-time monitoring and control of such important process variables as the solute concentration and the CSD (Yu *et al.*, 2004, 2007; Barrett *et al.*, 2005).

The PAT allows for establishing relations between the process actuators and the key process parameters. In this data-driven approach, the relations are derived on the basis of observations of changes in the monitored parameters caused by the applied actuator profiles. Effects of the crystallizer scale on the established relations and their cause can also be studied using PAT.

11.3.3
Measurement of the Solute Concentration

The use of attenuated total reflection Fourier transform infrared (ATR-FTIR) spectroscopy is reported by several researchers (see, for instance, Grön, Borissova, and Roberts, 2003). The main advantage of that technique is that it is applicable to multicomponent mixtures. This quality is especially important for crystallization of organic compounds and crystallization in the presence of impurities. Other real-time techniques such as, for instance, Raman spectroscopy (Tamagawa, Miranda, and Berglund, 2002), densitometry (Hojjati and Rohani, 2005), refractometry, and conductivity (Wilson *et al.*, 1991; Hložný, Sato, and Kubota, 1992) can be used as well.

The supersaturation can be inferred from the measured solute concentration and the equilibrium saturation concentration at the same temperature. The solubility is often correlated with the solution temperature by means of a simple mathematical expression in order to calculate the saturation concentration at any given temperature.

It is not sufficient to measure only the solute concentration in order to observe the process dynamics. A real-time measurement of the CSD is also required to establish a comprehensive relationship between the process variables and the measured data.

11.3.4
Measurement of the Crystal Number, Size, Distribution, and Morphology

A real-time measurement of the crystal attributes can be realized using online laser light-scattering techniques, in-line ultrasound probes, or in process focused beam reflectance measurement (FBRM) probe. The latter instrument has the advantage of being applicable to measurement in dense slurries without the need for sampling.

Image analysis is an emerging technique that has a potential for being used in monitoring and control of the crystal habit (Patience and Rawlings, 2001). A wide applicability of this technique is currently hindered by a poor image quality, difficulties with sampling or measuring in dense slurries, and problems with automatic image analysis for real-time extraction of the information on the crystal shape and size.

The use of specially designed sensors for measuring the fines content of the slurry is also reported (Rohani and Paine, 1987; Eek, 1995). This measurement is a prerequisite for feedback control of the CSD using fines removal and dissolution technique.

With progress in development and miniaturization of imaging techniques, real-time monitoring of the crystal morphology becomes possible (De Anda *et al.*, 2005; Patience, Dell'Orco, and Rawlings, 2004). However, automation of the sampling procedure and real-time image analysis for extracting the data on the crystal size and the CSD remain challenging tasks. Despite its immaturity, the first attempts to control the crystal shape using video microscopy are already reported (Patience and Rawlings, 2001).

Not only the sensor type, accuracy, and measurement frequency, but also the sensor location is crucial for the quality of information available for control. Distributed nature of crystallization poses problems in selection of a suitable location for the sensors. The sensors should be positioned in the place where the measurement is representative of the actual process performance, so that the changes in the measured process variables can be detected.

In the cases, when sensors for the primary variables such as the solute concentration or the CSD are not available, the measurements of the secondary variables can be used to reconstruct the values for the primary variables using the soft-sensing technology (see Chapters 13 and 14).

11.3.5
Actuators

A choice of the effective means of manipulating the process for reaching the posed objective is as critical as having the sensors for the key process variables. While recognizing importance of controlling the process, it has to be noted that process control is a complementary measure for obtaining an improved product quality. A proper process design is the primary measure for guaranteeing attainability of the desired product quality.

As the supersaturation is one of the most important process variables, methods of the supersaturation generation define available actuators. Depending on the type of the crystallization process, these actuators are the rate of cooling (in cooling crystallization), the rate of heating (in evaporative crystallization), and the rate of the antisolvent/reactant addition as well as the feed and solution concentrations (in antisolvent/reactive crystallization). If not determined in the process design phase on the basis of prevailing process conditions or process economics, location of antisolvent addition and type of antisolvent can be used as process parameters to influence various characteristics of the product crystals (see O'Grady *et al.*, 2007 and the references therein). Seeding is often used together with other methods of controlling the supersaturation to manage the initial desupersaturation profile, suppress nucleation, and influence the CSD of the product crystal population.

The agitation rate influences degree of mixing and suspension of solids in the crystallizer. The type of the used impeller and its frequency (profile) are important factors affecting the secondary nucleation rate especially for the systems with crystals prone to attrition or a narrow metastable zone.

The fines removal and fines dissolution rates are manipulated variables of the corrective nature. They are particularly effective in cases with unavoidable and uncontrolled nucleation. Fines dissolution can be exercised without separation and removal of fine crystals. The fines can be dissolved by temporarily creating undersaturation in the crystallizer by adding the solvent or raising the temperature.

The effectiveness of the available actuators is generally quite limited. Therefore, preventing the undesired effects from happening is certainly better than correcting for them. It is worth noting that an improper use of the process actuators can have detrimental effects similar to those caused by deliberate process disturbances. Cancelling out of the negative effects is usually not possible as many actuating techniques have an irreversible action. So, generators of the supersaturation are often not suitable for creation of undersaturation, and sources of nuclei cannot act as particle sinks.

11.4
Basic Recipe Control Strategy

The main strategy of the basic recipe control is to maintain a favorable balance between the crystal growth rate and the nucleation rate. The trade-off between crystal growth and nucleation is defined by the control objective that can be formulated in terms of the product quality, the process yield, or the batch time. For many crystallization systems in order to achieve the objective, it is sufficient to keep the supersaturation within the supersaturation region between the solubility line and the metastable limit in which no spontaneous nucleation occurs (the metastable zone) or within the limits justified either by the laboratory tests or data available in the process history database. Restricting the supersaturation to the defined limits helps to avoid or minimize nucleation, agglomeration, and an irregular crystal growth. The limits for the other actuators and parameters (parameters of the seeding procedure, impeller frequency, and fines removal rate) are defined to accomplish a similar effect of promoting crystal growth over nucleation.

Two approaches for the basic recipe control can be distinguished. The first one is the direct approach in which measurements of the process state variables (the solute concentration, the number of fines, and the CSD) are used for feedback control. The availability, the accuracy, and the frequency of the measurements are of the outmost importance in this case.

In the direct approach (Fujiwara et al., 2005), the supersaturation is controlled around a constant set-point or a set-point trajectory by feedback control. The numerical value of the set-point (trajectory) can be obtained through experimentation (Grön, Borissova, and Roberts, 2003). The supersaturation control at a constant level is a nearly optimal control policy whenever nucleation and agglomeration are negligible due to, for instance, an optimal seeding procedure applied to the process or the inherent properties of the system (Mullin, 2001).

For the systems dominated by the secondary nucleation, the set-point of the feedback controller is often not a constant but a complex function of the process variables and control objectives. Determination of a set-point supersaturation trajectory is not trivial in those cases. Therefore, instead of determination of a supersaturation profile, the indirect approach can be adopted for designing a control strategy. In this method, experiments are used to define optimal operating conditions and to derive optimized trajectories of the manipulated variables (the impeller frequency, the fines removal rate, the heat input, and the crystallizer temperature as functions of the process time or a state). The goal of such an empirical optimization procedure is to satisfy an objective consisting of requirements on the product CSD, the crystal size, or the fines content that are measured during experiments either off-line or preferably real-time using advanced measurement techniques.

11.4.1
How to Obtain a Recipe?

The first step in formulation of a basic recipe is determination of the operating zone of the supersaturation that is bounded by the solubility line and the metastable zone limit. The width of the metastable zone is variable and depends on many factors that influence spontaneous nucleation. There are different ways to determine the metastable zone width for a specific crystallization system (Tavare, 1987). Recent advances in measurement technology resulted in a growing number of publications reporting the use of ATR-FTIR spectroscopy (Fujiwara et al., 2002), FBRM measurement (Liotta and Sabesan, 2004), ultrasonic technique (Titiz-Sargut and Ulrich, 2002), and turbidimetry (Parsons, Black, and Colling, 2003) for detection of nucleation. The experimental procedure for determination of the metastable zone width can be fully automated (Liotta and Sabesan, 2004; Parsons, Black, and Colling, 2003).

Fujiwara et al. (2005) suggest using automated laboratory tests similar to those used for determination of the metastable zone width to obtain an optimized trajectory. The optimized trajectory is then a result of factorial experiment design studies that investigate the influence of process variables on the process performance (Togkalidou et al., 2001). The resulting operating policy is equipment-specific and might not be directly applicable to crystallizers that differ in configuration and scale. The main implication of the above disadvantage is that the findings of the laboratory tests can be used only as a starting point for operation at a larger scale. Hence, a similar process monitoring arrangement should be realized at a large scale to further optimize the process operation.

11.4.2
Scaling Up the Recipe

As discussed above, a recipe obtained at the laboratory scale might be not directly applicable to crystallizers of a larger scale. The recipe needs to be adjusted to obtain the control effect that was observed in the laboratory. The adjustment is required to compensate for differences in the crystallizer geometry, the crystallizer configuration, and hydrodynamic and mixing conditions.

There is no unified procedure on how to scale up the recipe. Ad-hoc approaches are usually used for each particular application. The optimization of the basic recipe at a larger scale is often performed experimentally. Some examples of the recipe scale-up are given in the subsequent sections.

Alternatively, scaling up of the recipe can be performed using inferential (Togkalidou et al., 2001), simplified (Kim and Lee, 2002; Doki et al., 1999), or elaborate (Kougoulos, Jones, and Wood-Kaczmar, 2006) modeling techniques providing an insight into the effects of the crystallizer scale on the relation between the process variables and the process performance measures.

11.5
Seeding as a Process Actuator

As a part of the supersaturation control, seeding is a powerful technique for manipulating the CSD and the supersaturation at the beginning of a batch (Chapter 12). It is mostly used to improve the (terminal-time as well as temporal) reproducibility of the product quality (Kalbasenka et al., 2007), the product purity, morphology, and the product CSD (Warstat and Ulrich, 2006). Seeding is often followed by one of the forms of the supersaturation control that can be either control at a constant supersaturation level found by the direct approach (Yu, Chow, and Tan, 2006), tracking of the input trajectory resulted from application of the indirect approach (Hojjati and Rohani, 2005), or an open-loop/a closed-loop optimal supersaturation control (Chapters 13 and 14).

Unseeded operation is prone to exhibit batch-to-batch variations due to the stochastic nature of the primary nucleation event (Kalbasenka et al., 2007). Seeding offers a possibility of managing the start-up of crystallization by influencing the initial growth and nucleation rates through a well-defined surface area of the seed crystals (Barrett et al., 2005). By tuning the seeding parameters (the mass, the size, the CSD of seed crystals, and the initial supersaturation), a low nucleation rate can be achieved throughout the batch (Doki et al., 2002).

The calculation procedures of Mullin (2001) and Doki et al. (2002) serve as a good starting point for choosing the right amount and size of seeds. However, those approached can be rather conservative as they might suggest a large mass of the seed crystals that can be economically prohibitive. Moreover, additional experiments are often required to achieve the desired quality of the end product (Yu, Chow, and Tan, 2006). The necessity of further experimental evaluation of the seeding effects is due to the fact that a simple correlation between the mass and the size of the product and the seed crystals does not include considerations on the seed quality and their growth behavior, the effect of other control policies such as the supersaturation control, and the crystallizer scale (Kalbasenka, 2009). Therefore, the seeding parameters and conditions should be optimized in connection with the system properties, scale, and control policies.

In Chapter 12, a set of heuristic rules is given for different systems. Next to the mass and the size of seed crystals, there are a number of additional parameters that should be carefully chosen to obtain a desired effect from seeding. They are discussed next.

11.5.1
Initial Supersaturation

The knowledge on the metastable zone width is essential for choosing an appropriate supersaturation point for seeding. On the one hand, if the supersaturation at seeding is higher or close to the metastability limit, a premature nucleation event or an irregular crystal growth can occur. Irregular crystals are known to promote nucleation due to their higher sensitivity to contact nucleation and breeding. A high initial supersaturation level can also be the cause of higher survival efficiency and the subsequent outgrowth of the nuclei generated after seeding or brought in with it.

On the other hand, a too low level of the supersaturation at seeding might be responsible for dissolution of seeds. This is often the case in evaporative crystallization processes especially when damaged crystals have a higher solubility and a slower growth rate.

The initial supersaturation is also a critical factor when relatively small seed loads are used. A small seed load might not deliver enough crystalline material to deplete supersaturation to the levels at which the secondary nucleation rate is negligibly low. The small load of seeds can be unintentionally created. If the seeds are not properly prepared for growth under the realistic operating conditions, they can partially dissolve after being fed into the crystallizer even in a supersaturated solution (Kalbasenka et al., 2007). The seed preparation is directly related to the seed quality as discussed below.

11.5.2
Seed Mass

The seed mass is proportional to the surface area of seed crystals and their size. Furthermore, in the ideal case, when the secondary nucleation is suppressed, the number of crystals introduced with seeding should result in the same number of product crystals. The condition of negligible nucleation is often met if a sufficiently large surface area of seeds is available for growth. Therefore, one should find an optimal combination of the seed size and the seed mass. A choice for a suitable combination of the seed size and mass depends on the process properties and the control objective. The suitable combination can be either found experimentally, chosen using the calculation procedure and the heuristic rules of Chapter 12 or calculated according to other approaches (see, for instance, (Doki *et al.*, 2002; Mersmann, 2001)).

11.5.3
Seed Size and Size Distribution

Ideally, the shape of the seed size distribution defines that of the product CSD given that the process is governed by the crystal growth and nucleation is negligible. Therefore, seeds with a narrow CSD are required for obtaining the product with a narrow unimodal product CSD. Narrow sieve fractions of milled or grown crystals are most frequently used as seeds.

To find an appropriate mass of seeds as described above, the CSD of the seed population is often approximated by a mean size. For practical reasons (tedious and irreproducible preparation, tendency to agglomeration, slow growth or even dissolution, and initial breeding), it is often not recommended to use the seeds with the size that is smaller than 100 μm. The seeds should also be not too large as the required seed mass can be uneconomically large with respect to the process yield. As reported by Heffels and Kind (1999), the seed mass used is usually between 0.1% and 3% of the total product mass (2–6% according to Lung-Somarriba *et al.*, 2004). Furthermore, large seed crystals can be the source of the secondary nucleation due to attrition. Lung-Somarriba *et al.* (2004) proposed that the seed size should be one quarter of the maximum achievable crystal size of the product to avoid attrition.

11.5.4
Seed Quality and Preparation Procedure

The seed preparation procedure should be economically attractive and reproducible. Milled seeds or fine crystals of the product from the previous batch are most frequently used due to relatively low cost of their production. Engineering of the seed crystals through growing them in well-defined conditions of the dedicated facilities is an alternative option. However, it is frequently not cost effective as it requires a similar or an even more complex measurement and control system than that of the full-scale installation.

The choice between milled seeds and the fines of the product crystals should be made based on the availability of the necessary amount of the required fraction after sieving the milled crystals or the product crystal mass. More often, however, the choice of the seed preparation method is dictated by the behavior of the obtained seed crystals (Aamir, Nagy, and Rielly, 2010). In some systems, milled crystals exhibit faster growth rates than their intact counterparts. In others, the stress induced onto the crystal lattice upon milling might lead to a retarded growth or even dissolution of crystals in supersaturated solutions. This is one of the reasons why aging of the seed crystals is employed (Kalbasenka et al., 2007). In the process of aging, the milled seeds are immersed in their own saturated solution. While in the solution, larger crystals are growing at the expense of dissolving smaller crystals. In other words, Ostwald ripening or coarsening of the crystals occurs (Mullin, 2001). The CSD of the crystal slurry is slowly approaching a narrow unimodal CSD. The crystals undergo healing, that is, the stress induced by milling is being released and the properties inherent to intact crystals (shape and growth behavior) are being restored. Another incentive for using aging is to dissolve fine crystals adhering to the seed particles and by doing so, avoid initial breeding. Alternatively, seeds can be washed to remove crystalline dust particles.

The origin of the seed crystals can also affect the product purity. According to Funakoshi, Takiyama, and Matsuoka (2001), the product purity was lower if milled crystals were seeded instead of the well-defined crystals of the same size. The observed decrease in purity was attributed to a higher secondary nucleation rate due to initial breeding. When milled seeds were used, the tendency to agglomeration was stronger and the crystal surface was rougher.

11.5.5
Methods of Addition of Seeds

The addition of seeds to the crystallizer should be performed using the methods that properly address transport, dosage, and dispersion of the crystalline material.

There are various ways of introducing the seeds. A well-defined mass of seeds can be added dry on the surface of liquid in the crystallizer. It is one of the simplest ways of seeding. Nevertheless, it is rarely recommended due to a poor dispersion, likely agglomeration, and initial breeding that nontreated seeds can induce.

To circumvent the problems mentioned above, the seeds are suspended in a saturated solution or a solvent and fed as slurry. Suspension ensures a better dispersion of crystals. A prolonged retention of seeds in the saturated solution favors dissolution of fines and ripening of the remaining crystals.

In the cases, when the surface roughness is a prerequisite for crystal growth, dry milled seeds suspended in solvent can be, however, preferred over slurry of well-defined crystals (Togkalidou et al., 2001).

The seed slurry should be fed to a place in the crystallizer where it can be rapidly propagated throughout the liquid volume. The potential candidates are the incoming streams (feed or recirculation streams if not undersaturated) and places near the crystallizer impeller.

11.6
Rate of Supersaturation Generation

Controlling the supersaturation within the metastable zone is the most common strategy for the basic recipe control. The direct control of the supersaturation at a constant level is often adopted. A constant set-point value of the supersaturation is tracked by a simple controller by manipulating the flow rate of a coolant in cooling crystallization, a heating agent in evaporative crystallization, and an antisolvent in antisolvent crystallization (Zhou et al., 2006). The process value of the supersaturation is calculated from the measurements of the solute concentration and the data on solubility. The concentration measurement is realized by means of a real-time measurement technique.

While using a constant supersaturation profile for control can result in an almost optimal operation for some crystallization systems (Mullin, 2001), derivation of supersaturation trajectories is needed for other systems to obtain a more optimal result. The optimized supersaturation trajectory can be obtained experimentally by refining the results of experiments with constant supersaturation profiles (Zhou et al., 2006). Multiple automated crystallizers can be employed in parallel to speed up the experimentation process.

The experimentally obtained supersaturation profile can be implemented on the process using proportional-integral (PI) controllers. Fujiwara et al. (2005) suggest using a state-dependent formulation of the optimized set-point profile instead of a time-dependent trajectory. In the former case, the supersaturation profile is a function of the solute concentration and a process (state) variable (such as temperature or the antisolvent addition rate) that serves as a set-point to the tracking controller. The advantage of using a state-dependent trajectory is a low sensitivity to process disturbances. In the latter case, the time-dependent process-variable trajectory that corresponds to the set-point supersaturation profile is applied to the process. Due to the process disturbances, discrepancies in the seeding conditions, and differences in the process conditions at different crystallizer scales, the deviations from the optimality observed in the laboratory can be significant.

According to Zhou et al. (2006), an easier tuning of the tracking controllers can be achieved if set-point of the supersaturation is transformed to the corresponding set-point for the controller of temperature or the antisolvent addition rate. In this way, the controller tuning becomes independent of the unknown crystallization kinetics.

The indirect approach for the supersaturation control can be applied whenever the real-time measurement of the supersaturation is not available. In that case, the process performance is optimized by adjusting the means of the supersaturation generation. For example, an optimized temperature profile that improves the CSD of the product during cooling batch crystallization can be obtained experimentally. Since there is an infinite number of possible profiles, calculations with a process model are often used to guide the search (Moscosa-Santilln et al., 2000).

As the logic suggests, the most optimal solution is using a combination of techniques that solve a variety of problems in an orderly fashion. So, seeding is applied to reduce inconsistency in the initial CSD, minimize secondary nucleation, and manage the initial desupersaturation profile. In order to reduce the mass of seeds needed, feedback control can assist in keeping the supersaturation within the metastable zone throughout the batch. In addition to the control techniques above, preventive and/or corrective measures are often required to minimize the secondary nucleation that is not effectively suppressed by the supersaturation control. For instance, the need for the CSD control often arises when contact nucleation is the dominant nucleation mechanism. In that case, stirring and fines removal rates can be manipulated to regulate the number of nuclei as described in the subsequent sections. Interactions among the applied control techniques should be managed in such a way, so that the desired balance between crystal growth and nucleation is achieved.

11.7
Mixing and Suspension of Solids

Success of the control policies described above depends on the degrees of mixing and suspension of solids in the crystallizer. In industrial crystallization, mechanical agitation is often used for mixing the slurry. Therefore, the stirring intensity is an important operating parameter as it influences micro-, meso-, and macromixing conditions.

A low stirring intensity can lead to a poor mixing that can result in temperature and concentration gradients in the crystallizer. As is shown by simulations using the compartmental modeling approach (Ma, Tafti, and Braatz, 2002), large concentration gradients can be responsible for different growth and nucleation rates at the top and the bottom of the crystallizer. Hence, application of the optimal recipe obtained in the laboratory conditions of nearly perfect mixing might show large deviations from the expected results at an industrial scale.

An insufficient mixing can also be responsible for variations in the suspension density and the CSD at different heights in the crystallizer (Sha et al., 1998). A low degree of mixedness can cause segregation and sedimentation of crystals and broadening of the CSD. The broad CSD is then a consequence of the internal classification leading to exclusion of large crystals from the growth process and an increase in the secondary nucleation rate due to a high concentration of large crystals in the vicinity of the impeller.

In antisolvent crystallization of the crystallization systems in which the crystal size is controlled by agglomeration and breakage, agitation rate has a significant influence on the particle formation processes. As reported by Shin and Kim (2002), the mean particle size reduces with an increased agitation speed.

In their study, O'Grady et al. (2007) showed that agitation intensity and antisolvent addition rate have influence on the metastable zone width and nucleation kinetics in the antisolvent crystallization of benzoic acid. The effect of agitation intensity was also shown to depend on the location of antisolvent addition.

Woo et al. (2006) performed simulation of a semibatch antisolvent crystallization of paracetamol from an acetone–water mixture using computational fluid dynamics (CFD). They found that higher stirring rates promote micromixing and the crystal growth rate by reducing mass transfer limitations on the crystal growth. A better mixing results in a faster desupersaturation and a lower nucleation rate. As a result, fewer and larger crystals are produced at higher agitation rates.

Kougoulos, Jones, and Wood-Kaczmar (2006) also used CFD calculations to analyze the influence of the impeller type and speed on the CSD and the process conditions during batch crystallization of an organic fine chemical on different laboratory scales. According to their findings, the commonly used scale-up rule of keeping a constant specific power input per unit mass can have a detrimental effect on the product quality at a larger scale despite an improved macromixing and a more homogeneous suspension of crystals. The product quality might degrade due to a less efficient heat transfer at lower local energy dissipation rates (poorer micromixing). Increasing the stirring intensity can improve heat transfer. However, this might result in an increased attrition rate and a larger fines content of the product CSD.

Contradictory findings of different researches stem from the differences in the system-specific growth and nucleation kinetics that lead to different trade-offs. Hence, an optimal stirring policy and its scaling up should be defined for each system individually.

From the above discussion, it is clear that for each particular application a trade-off has to be found between a sufficient homogeneity of the crystal suspension, good mixing, and mass transfer on the one hand and a moderate nucleation caused by agitation on the other hand. This trade-off yields a varying profile of the impeller frequency that results in an acceptable product quality. Determination of the acceptable impeller frequency trajectory can be done using real-time instruments that can monitor the changes in the CSD and the suspension density due to the varying stirring rate. The most simple control strategy, however, is to keep the stirrer at a constant rate that is found to produce the product with the best properties by, for instance, factorial design studies (Chianese, Di Cave, and Mazzarotta, 1984). This strategy can be justified for some systems as described in Section 13.6.

Generally speaking, a varying impeller frequency is often needed during batch crystallization due to the fact that the nucleation rate increases with accumulation of the solid mass. In the cases of a high sensitivity of the crystallized material to the secondary nucleation, a low-speed varying (Section 13.6) or even a low-speed intermittent stirring program should be applied to minimize attrition and preserve the crystal morphology. In the work of Iwamoto, Seki, and Koiwa (2002), the constant-speed recipe designed on a laboratory scale was applied to an industrial-scale crystallizer as an intermittent stirring program in which 5 min of stirring at a low constant speed were followed by 30 min of a nonagitation period.

Monitoring of the solid concentration and control of the suspension quality, the mean crystal size, and the CSD can be achieved using turbidimetry and manipulation of the stirring rate. Control of the crystal concentration was reported by Nomura, Shimizu, and Takahashi (2002). They made use of a multisensor measuring the slurry turbidity at different crystallizer heights. The reading from the phototransistor receiving the infrared light signal from the top emitting diode was compared to the average value calculated from the readings of all six phototransistors positioned along the crystallizer height. A proportional controller was used to minimize the obtained difference by adjusting the stirring rate. The CSD in the controlled run was narrower than that obtained with a constant impeller speed.

Apart from the indirect approaches described above, the direct control of the CSD using an FBRM probe can be realized by manipulating the stirring rate. The

FBRM measurement can assist in choosing an appropriate impeller frequency at which attrition of crystals is minimal (Kougoulos, Jones, and Wood-Kaczmar, 2005). Potentially, feedback control of the CSD could also be realized using the FBRM technique in the way similar to the CSD control by adjusting the fines removal rate (Tadayyon and Rohani, 2000). However, successful implementation of the feedback control strategy with the impeller frequency as an actuator requires unambiguous determination of the impeller frequency constraints. Real-time determination of the lower constraint is particularly challenging as the crystal size and the crystal mass increase with time. A measurement system similar to the one proposed by Nomura, Shimizu, and Takahashi (2002) could be used for real-time indication of the suspension quality.

The direct control of the mean particle size using turbidimetry is possible in the cases when an accurate relation between the mean crystal size and the slurry turbidity can be established (Raphael and Rohani, 1996). If the stirring speed is to be used as an actuator, the dependence of the crystal size on the stirring intensity should be studied.

If the used control strategies do not accomplish desired improvements in the product quality, changes to the hardware can be helpful. Various researchers showed that, for example, the impeller type can have a profound effect on the impact frequency of the crystal–impeller collisions, the flow pattern, the CSD (Shimizu, Nomura, and Takahashi, 1998; Kougoulos, Jones, and Wood-Kaczmar, 2006), and the mixedness in the crystallizer (Sha *et al.*, 1998). The impeller size and vertical position in the crystallizer are important parameters too (Shimizu, Nomura, and Takahashi, 1998). The hardness of the construction material of an impeller also has an influence on the secondary nucleation rate.

11.8
Fines Removal and Dissolution

Fines destruction is a remedial technique that is useful for correction of improper seeding and undesired nucleation. The availability of the remedial measures is particularly important for crystallization control due to an irreversible nature of the crystallization processes and limited possibilities for correcting the product quality in the postprocessing steps.

Fines removal and dissolution system can be an effective means of controlling the amount of fines in the systems with a progressive secondary nucleation (Jones, Chianese, and Mullin, 1984). Fines destruction can also be effective in correcting for improper seeding by removing the nuclei generated by initial breeding as well as in removing crystal fragments generated by the secondary nucleation at the later stage of the process. The presence of the fines dissolution system has a positive effect on the crystallizer performance since a narrower CSD with a larger median crystal size is usually attained in comparison to crystallization systems without the fines destruction (Kalbasenka *et al.*, 2006).

Rohani, Tavare, and Garside (1990) made use of an indirect measurement of the fines suspension density for the CSD control by manipulation of the fines removal rate. The temperature difference of the sampled slurry before and after dissolution of fines is related to the fines suspension density and therefore, it can be directly used for control (Rohani and Paine, 1987). Later, Tadayyon and Rohani (2000) employed a double-sensor turbidimeter as an indirect measurement of the fines suspension density to control the CSD.

The direct feedback control of the CSD became possible with appearance of real-time particle size analyzers. Randolph, Chen, and Tavana (1987) used a light-scattering instrument to infer and control the nuclei density in the fines loop by manipulating the fraction of the fines stream that was sent back to the crystallizer bypassing the fines dissolver.

Using FBRM measurement, Barthe and Rousseau (2006) monitored the changes in the CSD caused by fines removal and dissolution and obtained a recipe yielding crystals with a larger mean size. In the work of Tadayyon and Rohani (2000), an FBRM probe was already a part of a feedback control system. The FBRM signals corresponding to the count rate for chord lengths from 1 to 125 µm were utilized to control the number of fines by manipulating the fines withdrawal rate.

Dissolution of fines can also be realized without removing small crystals from the crystallizer and dissolving them in an external loop. The internal fines dissolution is achieved by creating undersaturation for short moments of time through, for instance, raising the temperature of a cooling batch crystallizer (Barrett *et al.*, 2005).

Doki *et al.* (2004) used the FBRM and ATR-FTIR measurements in the control algorithm to achieve a desired mean mass size by applying an alternating temperature profile in batch cooling crystallization. That is, they allowed heating in order to dissolve fines whenever the particle count number measured by FBRM increased above a predefined set-point value. The heating was stopped and the interrupted cooling program was resumed when the particle count reached its set-point. The heating was repeated as many times as it was necessary to keep the particle count constant. The cooling was stopped when the residual concentration measured by the ATR-FTIR probe was zero.

The major shortcoming of all corrective control techniques is that the secondary nuclei can be too small to be detected immediately. They become visible after they grow up to the detection limit of the used instrument. The introduced time delay can affect effectiveness of the control system in achieving a predefined CSD during the specified batch time.

11.9
Implementation of Basic Recipe Control

A distributed control system (DCS) or a programmable logic controller (PLC) system is a real-time control system that can be used to control batch manufacturing processes.

The logic behind the batch recipes introduced in the previous sections can be implemented within the DCS or the PLC system using, for instance, a sequential function chart (SFC). The SFC defines graphically the sequence of operations to be performed. The discrete states of the process are denoted as rectangles. The vertical lines connecting the states represent transitions between the states. The transitions occur if the corresponding conditions are met. Long horizontal single lines correspond to conditional paths (binary logic expression OR). Long horizontal double lines represent simultaneous execution of the paths (AND, see Figure 11.1b). In principle, PLCs can be used for all forms of proportional-integral-derivative (PID) control. However, in cases of more sophisticated control algorithms (inferential control or model-based control), a modern DCS or PLC–SCADA (supervisory control and data acquisition) system with the OPC (object linking and embedding for process control) functionality should be used. For more details on the programmable controllers and the ways of their programming, the reader is referred to Barker and Rawtani (2005) and Seborg, Edgar, and Mellichamp (2003).

In the simplified example of sequential logic control using an SFC that is illustrated in Figure 11.1, an operator initializes a seeded cooling batch crystallization process by turning on a manual switch. The crystallizer is then being filled with a saturated solution until the process value L_{pv} of the level indicator L (Figure 11.1a) reaches a predefined high level L_{sp_H}. Since the condition of $L_{pv} = L_{sp\_H}$ is true, a transition between the two process states occurs. Upon this transition, pump P-101 is stopped and valve V-101 is closed. The two next states, namely, agitation and conditioning, are activated by starting the stirrer motor M-1 and opening the coolant valve V-104.

Conditioning implies cooling of the solution to create the supersaturation. It is assumed here that the solute concentration can be measured directly and the supersaturation can be calculated online from the concentration measurement and solubility data. The predefined supersaturation level has a value that could be optimized to achieve an optimal effect from seeding. Conditioning proceeds until the actual supersaturation value reaches the set-point value at which seed addition should take place ($\sigma_{pv} = \sigma_{sp}$).

The seeds are added in the next step. The end of seeding is defined by the time needed to pump the seed slurry from the seeding vessel to the crystallizer. After the seed crystals are added, cooling continues. Manipulation of the coolant flow rate can either be used for following a temperature profile defined by the recipe or for the direct feedback control of the supersaturation.

The batch is terminated when a stopping criterion is met. The stopping criterion can be defined in terms of the product quality, the process yield, or the batch time. If the stopping criterion is met, cooling is interrupted by closing valve V-104. The crystallizer is then emptied by opening the discharge valve V-102 and starting the product pump P-102. When the level sensor L indicates that the level has reached its minimum value ($L_{pv} = L_{sp\_L}$), the two operations, namely agitation and discharge, are stopped. If cleaning is not required, the crystallization set-up is ready for the next batch.

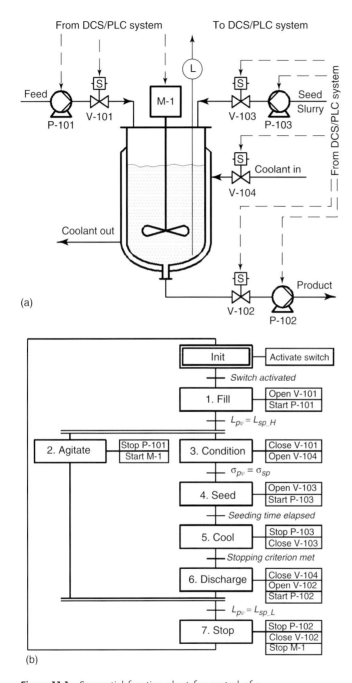

Figure 11.1 Sequential function chart for control of a seeded batch cooling crystallizer. (a) A seeded cooling batch crystallizer. (b) Example of an SFC.

The sequential logic control described above can include continuous control of process variables during a batch. In this way, the feedback control of, for instance, the crystallizer temperature or the supersaturation can be realized. Within a DCS system, the control of the process variables can be designed using control programs such as, for instance, a function block diagram (FBD). An FBD can contain different function blocks such as a PID controller. If desired, a time-varying set-point of the PID controller can be programmed in a sequence table that contains an experimentally found time profile of a control variable.

11.10
Conclusions

Under the urge of bringing a new product to the market as soon as possible, a basic recipe control strategy is a viable option for controlling a crystallization process. Based on experiments on a laboratory scale, a recipe can be devised and then scaled up to the production scale. A simple controller can be used to implement the recipe.

The disadvantage of both the direct and the indirect design approaches for crystallization control lies in their empirical nature. The experimental studies may require a large number of experiments to design and optimize the recipe. Scaling-up of the experimentally obtained recipe is not straightforward. Therefore, the performance of the recipe at a larger scale can be far from optimal.

Owing to its simplicity, feedback control using direct or indirect measurement of the key process variables is easy to implement. However, performance of such a control system is highly dependent on the used measurement technique.

In order to improve on the process performance, a more profound insight into the crystallization mechanisms and their interactions is needed. The model-based optimization and control techniques introduced in Chapters 13 and 14 offer a possibility for getting a better understanding and control of the batch crystallization processes.

References

Aamir, E., Nagy, Z.K., and Rielly, C.D. (2010) Evaluation of the effect of seed preparation method on the product crystal size distribution for batch cooling crystallization processes. *Cryst. Growth Des.*, **10**, 4728–4740.

De Anda, J.C., Wang, X.Z., Lai, X., Roberts, K.J., Jennings, K.H., Wilkinson, M.J., Watson, D., and Roberts, D. (2005) Real-time product morphology monitoring in crystallization using imaging technique. *AIChE J.*, **51** (5), 1406–1414.

Bakeev, K. (ed.) (2005) *Process Analytical Technology. Spectroscopic Tools and Implementation Strategies for the Chemical and Pharmaceutical Industries*, Blackwell Publishing Ltd, Oxford.

Barker, M. and Rawtani, J. (2005) *Practical Batch Process Management*, Newnes, Oxford.

Barrett, P., Smith, B., Worlitschek, J., Bracken, V., O'Sullivan, B., and O'Grady, D. (2005) A review of the use of process analytical technology for the understanding and optimization of production batch crystallization processes. *Org. Process Res. Dev.*, **9** (3), 348–355.

Barthe, S. and Rousseau, R. (2006) Utilization of focused beam reflectance measurement in the control of crystal size distribution in a batch cooled crystallizer. *Chem. Eng. Technol.*, **29** (2), 206–211.

Chianese, A., Di Cave, S., and Mazzarotta, B. (1984) Investigation on some operating factors influencing batch cooling crystallization, in *Industrial Crystallization 84*, (eds S. Jančić and E. de Jong), Proceedings of the 9th Symposium on Industrial Crystallization, September 25–28, Elsevier, Amsterdam, The Netherlands, pp. 443–446.

Doki, N., Hamada, O., Kubota, N., Masumi, F., Sato, A., and Yokota, M. (1999) Scale-up experiments on seeded batch cooling crystallization of potassium alum. *AIChE J.*, **45** (12), 2527–2533.

Doki, N., Kubota, N., Yokota, M., and Chianese, A. (2002) Determination of critical seed loading ratio for the production of crystals of uni-modal size distribution in batch cooling crystallization of potassium alum. *J. Chem. Eng. Jpn.*, **35**, 670–676.

Doki, N., Seki, H., Takano, K., Asatani, H., Yokota, M., and Kubota, N. (2004) Process control of seeded batch cooling crystallization of the metastable α-form glycine using an *in-situ* ATR-FTIR spectrometer and an *in-situ* FBRM particle counter. *Cryst. Growth Des.*, **4** (5), 949–953.

Eek, R. (1995) Control and dynamic modelling of industrial suspension crystallizers. PhD thesis. Delft University of Technology, The Netherlands.

Fujiwara, M., Chow, P., Ma, D., and Braatz, R. (2002) Paracetamol crystallization using laser backscattering and ATR-FTIR spectroscopy: metastability, agglomeration, and control. *Cryst. Growth Des.*, **2** (5), 363–370.

Fujiwara, M., Nagy, Z., Chew, J., and Braatz, R. (2005) First-principles and direct design approaches for the control of pharmaceutical crystallization. *J. Process Control*, **15**, 493–504.

Funakoshi, K., Takiyama, H., and Matsuoka, M. (2001) Influences of seed crystals on agglomeration phenomena and product purity of m-chloronitrobenzene crystals in batch crystallization. *Chem. Eng. J.*, **81**, 307–312.

Grön, H., Borissova, A., and Roberts, K. (2003) In-process ATR-FTIR spectroscopy for closed-loop supersaturation control of a batch crystallizer producing monosodium glutamate crystals of defined size. *Ind. Eng. Chem. Res.*, **41** (1), 198–206.

Heffels, S. and Kind, M. (1999) Seeding technology: an underestimated critical success factor for crystallization. Proceedings of the 14th International Symposium on Industrial Crystallization, September 12–16, Cambridge, United Kingdom.

Hložný, L., Sato, A., and Kubota, N. (1992) On-line measurement of supersaturation during batch cooling crystallization of ammonium alum. *J. Chem. Eng. Jpn.*, **25** (5), 604–606.

Hojjati, H. and Rohani, S. (2005) Cooling and seeding effect on supersaturation and final crystal size distribution (CSD) of ammonium sulphate in a batch crystallizer. *Chem. Eng. Process.*, **44**, 949–957.

Iwamoto, K., Seki, M., and Koiwa, Y. (2002) Batch cooling crystallization of an organic photochemical with a low-speed intermittent stirring technique. *J. Chem. Eng. Jpn.*, **35** (11), 1105–1107.

Jones, A., Chianese, A., and Mullin, J. (1984) Effect of fines destruction on batch cooling crystallization of potassium sulphate solutions, in *Industrial Crystallization 84* (eds S. Jančić, and E. de Jong), Proceedings of the 9th Symposium on Industrial Crystallization, September 25–28, The Hague, Elsevier, Amsterdam, The Netherlands, pp. 191–195.

Kalbasenka, A.N. (2009) Model-based control of industrial batch crystallizers: experiments on enhanced controllability by seeding actuation. PhD thesis. Delft University of Technology, The Netherlands. ISBN: 978-90-6464-361-3, *www.dcsc.tudelft.nl/Research/PhDtheses/*.

Kalbasenka, A., Spierings, L., Huesman, A., and Kramer, H. (2006) Effect of crystallizer scale and actuator configuration on the product quality in a seeded fed-batch crystallization of ammonium sulphate, in *The 13th International Workshop on Industrial Crystallization* (eds P. Jansens, J. ter Horst, and S. Jiang), IOS Press, Delft,

The Netherlands, pp. 138–145, September 13–15.

Kalbasenka, A., Spierings, L., Huesman, A., and Kramer, H. (2007) Application of seeding as a process actuator in a model predictive control framework for fed-batch crystallization of ammonium sulphate. *Part. Part. Syst. Charact.*, **24** (1), 40–48.

Kim, K.-J. and Lee, C.-H. (2002) Scale-up study on unseeded batch cooling crystallization using co-solvent. *J. Chem. Eng. Jpn.*, **35** (11), 1091–1098.

Kim, S., Lotz, B., Lindrud, M., Girard, K., Moore, T., Nagarajan, K., Alvarez, M., Lee, T., Nikfar, F., Davidovich, M., Srivastava, S., and Kiang, S. (2005) Control of the particle properties of a drug substance by crystallization engineering and the effect on drug product formulation. *Org. Process Res. Dev.*, **9** (6), 894–901.

Kougoulos, E., Jones, A., and Wood-Kaczmar, M. (2005) Modelling particle disruption of an organic fine chemical compound using Lasentec focused beam reflectance monitoring (FBRM) in agitated suspensions. *Powder Technol.*, **155**, 153–158.

Kougoulos, E., Jones, A., and Wood-Kaczmar, M. (2006) Process modelling tools for continuous and batch organic crystallization processes including application to scale-up. *Org. Process Res. Dev.*, **10** (4), 739–750.

Liotta, V. and Sabesan, V. (2004) Monitoring and feedback control of supersaturation using ATR-FTIR to produce an active pharmaceutical ingredient of a desired crystal size. *Org. Process Res. Dev.*, **8** (3), 488–494.

Lung-Somarriba, B., Moscosa-Santillán, M., Porte, C., and Delacroix, A. (2004) Effect of seeded surface area on crystal size distribution in glycine batch cooling crystallization: a seeding methodology. *J. Cryst. Growth*, **270** (3–4), 624–632.

Ma, D., Tafti, D., and Braatz, R. (2002) Optimal control and simulation of multidimensional crystallization processes. *Comput. Chem. Eng.*, **26**, 1103–1116.

Mersmann, A. (ed.) (2001) *Crystallization Technology Handbook*, 2nd edn, Marcel Dekker, New York.

Moscosa-Santillán, M., Bals, O., Fauduet, H., Porte, C., and Delacroix, A. (2000) Study of batch crystallization and determination of an alternative temperature-time profile by on-line turbidity analysis–application to glycine crystallization. *Chem. Eng. Sci.*, **55**, 3759–3770.

Mullin, J. (2001) *Crystallization*, 4th edn, Butterworth-Heinemann, Oxford.

Nomura, T., Shimizu, K., and Takahashi, K. (2002) Control of crystal size distribution in a seeded batch crystallizer. *J. Chem. Eng. Jpn.*, **35**, 1108–1112.

O'Grady, D., Barrett, M., Casey, E., and Glennon, B. (2007) The effect of mixing on the metastable zone width and nucleation kinetics in the anti-solvent crystallization of benzoic acid. *Trans. IChemE, Part A, Chem. Eng. Res. Des.*, **85** (A7), 945–952.

Parsons, A., Black, S., and Colling, R. (2003) Automated measurement of metastable zones for pharmaceutical compounds. *Trans. IChemE.*, **81A**, 700–704.

Patience, D., Dell'Orco, P., and Rawlings, J. (2004) Optimal operation of a seeded pharmaceutical crystallization with growth-dependent dispersion. *Org. Process Res. Dev.*, **8** (4), 609–615.

Patience, D. and Rawlings, J. (2001) Particle-shape monitoring and control in crystallization processes. *AIChE J.*, **47** (9), 2125–2130.

Randolph, A., Chen, L., and Tavana, A. (1987) Feedback control of CSD in a KCl crystallizer with a fines dissolver. *AIChE J.*, **33** (4), 583–591.

Raphael, M., and Rohani, S. (1996) On-line estimation of solids concentrations and mean particle size using a turbidimetry method. *Powder Technol.*, **89**, 157–163.

Rohani, S. and Paine, K. (1987) Measurement of solid concentration of a soluble compound in a saturated slurry. *Can. J. Chem. Eng.*, **65**, 163–165.

Rohani, S., Tavare, N., and Garside, J. (1990) Control of crystal size distribution in a batch cooling crystallizer. *Can. J. Chem. Eng.*, **68**, 260–267.

Seborg, D.E., Edgar, T.E., and Mellichamp, D.A. (2003) *Process Dynamics and Control*, 2nd edn, John Wiley & Sons, Inc., New York.

Sha, Z., Louhi-Kultanen, M., Ogawa, K., and Palosaari, S. (1998) The effect of mixedness on crystal size distribution in a continuous crystallizer. *J. Chem. Eng. Jpn.*, **31** (1), 55–60.

Shimizu, K., Nomura, T., and Takahashi, K. (1998) Crystal size distribution of aluminum potassium sulfate in a batch crystallizer equipped with different types of impeller. *J. Cryst. Growth*, **191**, 178–184.

Shin, D.-M. and Kim, W.-S. (2002) Drowning-out crystallization of L-ornithine-aspartate in turbulent agitated reactor. *J. Chem. Eng. Jpn.*, **35** (11), 1083–1090.

Tadayyon, A. and Rohani, S. (2000) Control of fines suspension density in the fines loop of a continuous KCl crystallizer using transmittance measurement and an FBRM probe. *Can. J. Chem. Eng.*, **78**, 633–673.

Tamagawa, R.E., Miranda, E.A., and Berglund, K.A. (2002) Raman spectroscopic monitoring and control of aprotinin supersaturation in hanging-drop crystallization. *Cryst. Growth Des.*, **2** (4), 263–267.

Tavare, N. (1987) Batch crystallization: a review. *Chem. Eng. Commun.*, **61**, 259–318.

Titiz-Sargut, S. and Ulrich, J. (2002) Influence of additives on the width of the metastable zone. *Cryst. Growth Des.*, **2** (5), 371–374.

Togkalidou, T., Braatz, R., Johnson, B., Davidson, O., and Andrews, A. (2001) Experimental design and inferential modeling in pharmaceutical crystallization. *AIChE J.*, **47** (1), 160–168.

Warstat, A. and Ulrich, J. (2006) Seeding during batch cooling crystallization–an initial approach to heuristic rules. *Chem. Eng. Technol.*, **29** (2), 187–190.

Wilson, D., Lee, P., White, E., and Newell, R. (1991) Advanced control of a sugar crystallizer. *J. Process Control*, **1** (4), 197–206.

Woo, X., Tan, R., Chow, P., and Braatz, R. (2006) Simulation of mixing effects in antisolvent crystallization using a coupled CFD-PDF-PBE approach. *Cryst. Growth Des.*, **6** (6), 1291–1303.

Yu, L., Lionberger, R., Raw, A., D'Costa, R., Wu, H., and Hussain, A. (2004) Applications of process analytical technology to crystallization processes. *Adv. Drug Delivery Rev.*, **56**, 349–369.

Yu, Z., Chew, J., Chow, P., and Tan, R. (2007) Recent advances in crystallization control. An industrial perspective. *Chem. Eng. Res. Des.*, **85** (A7), 893–905.

Yu, Z., Chow, P., and Tan, R. (2006) Seeding and constant-supersaturation control by ATR-FTIR in anti-solvent crystallization. *Org. Process Res. Dev.*, **10** (4), 717–722.

Zhou, G., Fujiwara, M., Woo, X., Rusli, E., Tung, H., Starbuck, C., Davidson, O., Ge, Z., and Braatz, R. (2006) Direct design of pharmaceutical antisolvent crystallization through concentration control. *Cryst. Growth Des.*, **6** (4), 892–898.

12
Seeding Technique in Batch Crystallization

Joachim Ulrich and Matthew J. Jones

12.1
Introduction

Seeding is a commonly applied procedure for initiating a crystallization process in a reproducible manner and describes the technique of adding crystalline (seed) material once the initial supersaturation has been defined by the usual means and prior to the occurrence of nucleation. The addition of the crystal seeds initiates the crystallization process by inducing growth of the seed crystals due to the supersaturation of the solution (or, conversely, by reducing supersaturation by allowing the seed crystals to grow). The purpose of seeding is to provide surface area for crystals to grow without having to rely upon nucleation. The level of supersaturation, together with the amount of available crystal surface area provided by the seeds, defines both the growth rate of the crystals and their quality (since ''quality'' depends on growth rate). The crystal size distribution of the final product depends on the nature (size distribution) of the seeds introduced into the process (Heffels and Kind, 1999; Kind, 2004, Kubota, 2001; Kubota *et al.*, 2002; Mullin, 1993; Warstat and Ulrich, 2003; Zhang and Grant, 2005).

12.2
Seeding Operation: Main Principles and Phenomena

Most crystalline products sold in the market have to meet specific and often high demands regarding product quality, for example, purity, mean crystal size, and size distribution. As a consequence, industrial processes are required to deliver the product with reliable and reproducible crystal quality that meets these specifications. This can be achieved by employing a defined and reproducible crystal growth rate at that point of the process where the seeds are added. In order to ensure that the required conditions are met, temperature and concentration need to be known. These properties can be determined online using sound velocity measurements.

Industrial Crystallization Process Monitoring and Control, First Edition. Edited by
Angelo Chianese and Herman J. M. Kramer.
© 2012 Wiley-VCH Verlag GmbH & Co. KGaA. Published 2012 by Wiley-VCH Verlag GmbH & Co. KGaA.

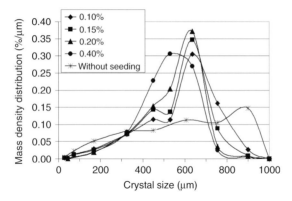

Figure 12.1 Mass density distribution of product crystals for different amounts of seed crystals (Warstat and Ulrich, 2003). The seed amount is given as percentage mass relative to the mass of the solute.

However, growth rate is not the only factor that determines the reproducibility of a crystal growth process. The necessary surface area for growth provided by adding crystal seeds to the process is a function of seed size and the number of utilized seeds. The quality of the surface area of the seeds is also important and depends to some extent on the history of the seed crystals and the way the seeds are added (as a dry powder or as a wet suspension). If the surface area provided is insufficient, the supersaturation that is continuously created as a result of the temperature gradient applied, solvent evaporation, or the addition of an (anti)solvent will increase at a greater rate than that at which it can be removed by virtue of the crystal growth rate of the seeds. As a result, the supersaturation will increase and heterogeneous nucleation can occur, resulting in, at a minimum, a bimodal or multimodal crystal size distribution (Figure 12.1 data "without seeding" and 0.1 and 0.15% seed). This type of distribution is generally undesirable with respect to the product size distribution, since the fines can result in dust or facilitate caking and, in addition, can be hazardous.

A good quality crystalline product with a well-defined crystal size and size distribution can be achieved if certain criteria are met during the process. An extensive study by Warstat and Ulrich (2006, 2007) and Warstat (2006) resulted in clear rules on how to approach such a seeding procedure.

12.3
Use of Seeding for Batch Crystallization: Main Process Parameters

A number of key parameters have to be determined and defined and, thereafter, a few simple experiments can provide a clear picture on how to approach a particular crystallization process. The recommended procedure is discussed in the following text, by means of a few examples employing cooling crystallization.

In a batch crystallization process, the type of temperature profile applied in the course of the process, or rather the supersaturation enforced by means of this temperature profile, has a significant impact on the resulting crystalline product. Three types of temperature profiles are commonly used: natural cooling, a linear cooling profile, or a parabolic temperature profile (Hu *et al.*, 2005; Costa and Maciel Filho, 2005; Mohamed, Abu-Jdayil, and Al Khateeb, 2002; Miller and Rawlings, 1994; Nyvlt, 1988; Ulrich, 1979; Mullin and Nyvlt, 1971; Jones, 1974). As the increase in solid surface area available due to crystal growth is parabolic, the change in supersaturation should, ideally, also follow a parabolic function. At the start of the crystallization process, the change in temperature, and thus the supersaturation imposed, should be slow and commensurate with the small surface area provided by the crystal seeds. As the crystals grow, the rate at which the temperature is changed should increase, with the most rapid rate change achieved toward the end of the process when the grown crystals have the largest surface area. When properly controlled, a parabolic temperature change results in a constant crystal growth rate throughout the duration of the process (see Figures 12.2 and 12.3), which in turn is conducive to obtaining a good quality crystalline product. As the calculation of the temperature profile is quite involved, a simplified approach as proposed by Mullin (1993) (Equation (12.1)) is employed in most cases, leading to a temperature profile with a cubic time dependence.

$$T(t) = T_0 - (T_0 - T_E)\left(\frac{t}{\tau}\right)^3 \tag{12.1}$$

The temperature–time profile $T(t)$ depends on the initial temperature T_0, the final temperature T_E at time t and on the retention time of the process, τ. In the following, the expression "controlled cooling" implies this type of cooling profile.

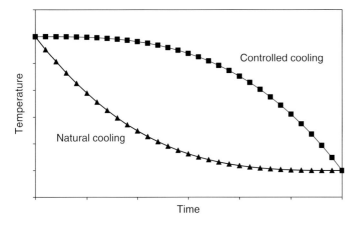

Figure 12.2 Temperature profiles for natural cooling and controlled cooling in batch crystallization (Ulrich and Glade, 1999).

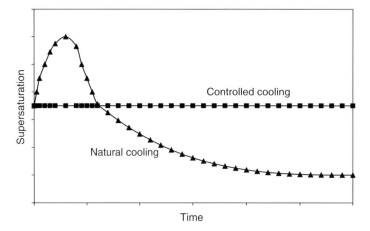

Figure 12.3 Supersaturation profiles for natural cooling and controlled cooling batch crystallization (Ulrich and Glade, 1999).

In the case of natural (Newtonian) cooling, the initial change in temperature and, by extension, supersaturation is very rapid and usually leads to spontaneous nucleation and a bimodal crystal size distribution (Figure 12.1 without seeding).

Figure 12.4 shows how cooling profile and seeding can influence product crystal purity in a particular system, in this case an organic salt (Warstat and Ulrich, 2003).

For all three distributions shown, linear cooling without seeding results in the highest concentration of impurities, whereas the combination of controlled (parabolic) cooling and seeding significantly reduces the amount of impurities present in the product.

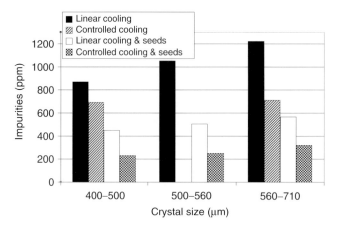

Figure 12.4 Impurity contents in the crystals obtained from different crystallization procedures (different cooling profiles with and without seeding) for different crystal size ranges (Warstat and Ulrich, 2003).

12.4
Control of Batch Crystallization by Seeding: Empirical Rules for Design

Any other constraints upon the process conditions will depend on the physical properties of both the product and the liquor from which it is crystallized. In particular, the properties that need to be considered are

- the width of the metastable zone (MSZ);
- the maximum growth rate acceptable with respect to product quality;
- the propensity of the solid to agglomerate; and
- the tendency of the system toward secondary nucleation.

On the basis of a detailed experimental study of five substances (critic acid, K_2SO_4, $NaNO_3$, adipic acid, and an organic heterocyclic salt) Warstat and Ulrich (2006, 2007) and Warstat (2006) defined ranges of these properties that were then considered to represent either high or low values:

	Small/low	Large/high
Metastable zone	0–5 K	~10 K
Growth rate	10^{-9} m/s	10^{-8} m/s
Agglomeration tendency	$N_s/N_p < 1$	$N_s/N_p > 1$
Secondary nucleation	$N_s/N_p > 1$	$N_s/N_p < 1$

The agglomeration tendency was defined by the ratio of the number of seed crystals N_s to the number of product crystals N_p according to Derenzo, Shimizu, and Giulietti (1996).

The quantification of the tendency toward a lesser or greater extent of secondary nucleation was based on the same considerations. With the assumption that the product crystal size $L_p = L_{50}$, the mass increase can be calculated from a knowledge of the supersaturation. Monitoring the number of product crystals allows the propensity of the system toward secondary nucleation or agglomeration to be determined. An increase in the number of particles indicates a tendency toward secondary nucleation. Conversely, a reduction in the number of particles shows a tendency toward agglomeration. In an ideal process without secondary nucleation or agglomeration, the number of particles stays constant and the ratio $N_s/N_p = 1$.

Providing the behavior of the system, with respect to the properties above, is known and the temperature profile driving the supersaturation is defined, simple guidelines can be defined that can be used as an aid to achieving a particular, desired outcome of the crystallization process. The resulting product obtained in the process can be defined either in terms of a product with maximum purity and narrow, unimodal particle size distribution (case A), in terms of a product with a narrow size distribution and a specific, predefined size (case B), or merely a narrow

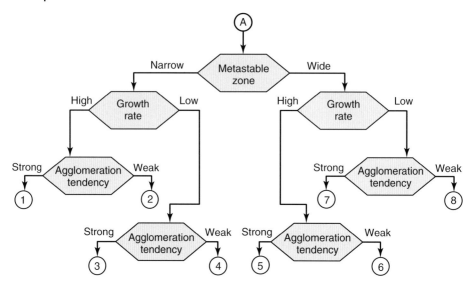

Figure 12.5 Flowchart detailing queries to the crystallization system and resulting rules for controlling the crystallization process in the case where the desired product should have both a unimodal particle size distribution and high purity.

size distribution (case C). For each of these three cases, the following flow charts depict the rules that need to be obeyed, in order to achieve the particular objective.

The seed mass referred to in the following equation is based on the mass of seed crystals M_s of size L_s required to achieve a particular product crystal mass M_p with size L_p in accordance with the equation provided by Mullin (1993) (Equation (12.2)):

$$M_s = M_p \frac{L_s^3}{(L_p^3 - L_s^3)} \tag{12.2}$$

For case A, where a high product purity is demanded in addition to a unimodal particle size distribution, the system has to be queried according to the structure provided in the decision chart shown in Figure 12.5. A low degree of agglomeration is helpful in achieving high purity, as processes in which particles display a tendency toward pronounced agglomerate formation are frequently found to be impure (Funakoshi, Takiyama, and Matsuoka, 2000). For this reason, the system has to be classified in accordance with its propensity for agglomerate formation. In this case, a total of (1–8) rules are given regarding how best to control a crystallization process using seeding.

- **Rule 1 (narrow MSZ; high growth rate; high propensity for agglomeration):**
  - large seed crystals should be employed;
  - seed mass added should be greater than the mass suggested by Equation (12.2), controlled cooling; and

- due to the tendency of the particles to agglomerate, M_s should not be selected much above that given by Equation (12.2).
- **Rule 2 (narrow MSZ; high growth rate; low propensity for agglomeration)**:
  - seed mass added should be greater than the mass suggested by Equation (12.2), controlled cooling;
  - very high seed mass in order to avoid dependence upon cooling rate.
- **Rule 3 (narrow MSZ; small growth rate; high propensity for agglomeration)**:
  - combination of seeding and parabolic cooling rate;
  - seed mass added should be greater than the mass suggested by Equation (12.2), controlled cooling;
  - due to the tendency of the particles to agglomerate, M_s should not be selected much above that given by Equation (12.2); and
  - large seed crystals preferably to be used.
- **Rule 4 (narrow MSZ; small growth rate; low propensity for agglomeration)**:
  - seed mass added should be greater than the mass suggested by Equation (12.2), controlled cooling.
- **Rule 5 (wide MSZ; high growth rate; high propensity toward agglomeration)**:
  - large seed crystals preferably to be used;
  - unimodal particle size distribution achievable with controlled cooling and seed mass according to Equation (12.2);
  - higher seed mass leads to narrower particle size distribution independent of cooling rate; and
  - to avoid agglomeration, seed mass should not be much greater than the mass suggested by Equation (12.2).
- **Rule 6 (wide MSZ; high growth rate; low propensity toward agglomeration)**:
  - unimodal particle size distribution achievable with controlled cooling and seed mass according to Equation (12.2);
  - higher seed mass leads to narrower particle size distribution independent of cooling rate.
- **Rule 7 (wide MSZ; small growth rate; high probability of agglomeration)**:
  - large seed crystals preferably to be used;
  - moderate seed mass to avoid agglomeration;
  - seeding combined with controlled cooling; and
  - seed mass according to Equation (12.2), if the growth rate is extremely low, higher seed mass may be beneficial.
- **Rule 8 (wide MSZ; low growth rate; low probability of agglomeration)**:
  - seed mass used should be higher than that calculated with Equation (12.2);
  - if a lower seed mass is used, controlled cooling should be applied; and
  - a very high seed mass leads to a particle size distribution that is independent of the cooling rate.

For the second case B, where a unimodal particle size distribution is desired together with a specific mean particle size, the corresponding decision tree is shown in Figure 12.6. In order to achieve a particular mean particle size, it is important to avoid attrition, as the secondary nuclei will affect the width and the

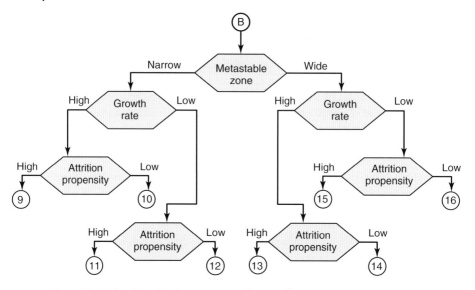

Figure 12.6 Flowchart detailing queries to the crystallization system and resulting rules for controlling the crystallization process in the case where the desired product should have both a unimodal particle size distribution and a particular mean size.

mean of the particle size distribution. A further set of eight rules (number 9–16) is provided.

- **Rule 9 (narrow MSZ; high growth rate; high attrition)**:
  - final mean particle size should not be too large;
  - seed crystals should be as small as possible;
  - seeding combined with controlled cooling; and
  - seed mass greater than calculated with Equation (12.2).
- **Rule 10 (narrow MSZ; high growth rate; low attrition)**:
  - seeding coupled with parabolic cooling rate;
  - higher seed mass than the mass calculated with Equation (12.2); and
  - very high seed mass results in independence from cooling rate and reduces the mean particle size.
- **Rule 11 (narrow MSZ; low growth rate; high attrition)**:
  - final mean particle size should not be too large;
  - seed crystals should be as small as possible;
  - seeding coupled with controlled cooling; and
  - seed mass employed should be greater than calculated with Equation (12.2).
- **Rule 12 (narrow MSZ; low growth rate; low attrition)**:
  - seeding coupled with controlled cooling;
  - seed mass employed should be greater than calculated with Equation (12.2); and

12.4 Control of Batch Crystallization by Seeding: Empirical Rules for Design | 135

- in order not to unduly restrict the mean particle size, the seed mass should not be too high.
- **Rule 13 (wide MSZ; high growth rate; high attrition):**
  - final mean particle size should not be too large;
  - seed crystals should be as small as possible;
  - unimodal particle size distribution can be achieved with controlled cooling using a seed mass in accordance with Equation (12.2); and
  - higher seed mass results in a narrower particle size distribution independent of the cooling rate as well as reducing the mean particle size.
- **Rule 14 (wide MSZ; high growth rate; low attrition):**
  - unimodal particle size distribution can be achieved with controlled cooling using a seed mass in accordance with Equation (12.2);
  - higher seed mass results in a narrower particle size distribution independent of the cooling rate; and
  - seed mass should not be too high, as otherwise the mean particle size will be reduced.
- **Rule 15 (wide MSZ; low growth rate; high attrition):**
  - final mean particle size should not be too high;
  - seed crystals should be as small as possible;
  - seed mass should be higher than that given by Equation (12.2);
  - for small seed mass, seeding should be combined with controlled cooling; and
  - higher seed mass results in a narrower particle size distribution independent of the cooling rate as well as reducing the mean particle size.
- **Rule 16 (wide MSZ; low growth rate; low attrition):**
  - seed mass employed should be greater than that calculated with Equation (12.2);
  - for small seed mass seeding should be combined with controlled cooling; and
  - higher seed mass results in a narrower particle size distribution independent of cooling rate as well as reducing the mean particle size.

In those cases where a unimodal particle size distribution is required (case C), the decision tree shown in Figure 12.7 should be employed. In this case, the main requirement is to avoid secondary nucleation. For this reason, the flow chart queries only the MSZ width of the system and the growth rate of the crystals. Attrition and agglomeration have to be considered in this case, but are secondary to the above factors. Multimodal particle size distributions must be expected in those situations where, in the presence of crystals, secondary nucleation cannot be avoided. Should this situation arise, the process can nonetheless be optimization of by means of the remaining criteria regarding attrition and agglomeration (rules 1–16). Four additional rules (17–20) can be stated for case C.

- **Rule 17 (narrow MSZ; high growth rate):**
  - seeding coupled with controlled cooling;
  - seed mass employed should be higher than that calculated with Equation (12.2); and

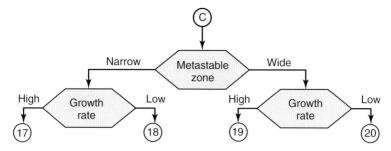

Figure 12.7 Flowchart detailing queries to the crystallization system and resulting rules for controlling the crystallization process in the case where the only product quality required is a unimodal particle size distribution.

- higher seed mass yields a narrower particle size distribution independent of the cooling rate.
- **Rule 18 (narrow MSZ; low growth rate):**
  - seeding combined with controlled cooling;
  - seed mass should be greater than that calculated with Equation (12.2).
- **Rule 19 (broad MSZ; high growth rate):**
  - unimodal particle size distribution can be achieved with a seed mass according to Equation (12.2) and by employing controlled cooling;
  - increasing the seed mass results in a narrower particle size distribution independent of cooling rate.
- **Rule 20 (broad MSZ; low growth rate):**
  - seed mass used should exceed that calculated with Equation (12.2);
  - for lower seed masses, seeding should be combined with controlled cooling; and
  - increasing the seed mass results in a narrower particle size distribution independent of cooling rate.

The heuristic given above is a set of 20 empirical rules derived from experimental observations. The objective of the experiments from which these rules were derived was to optimize the seeded batch crystallization of five different systems (potassium sulfate, sodium nitrate, adipic acid, citric acid, and a heterocyclic organic salt), selected with a view to covering a broad range of particle properties, with respect to the criteria that form the basis of the three cases described above. The insights gathered from these experiments were combined with existing knowledge on the nucleation and growth behavior of the compounds and summarized in the heuristic rules presented above. These rules aid the selection of near-optimum process conditions for the specific purpose of generating defined product properties (purity, particle size distribution, and mean particle size). The following additional rules should be adhered to as a matter of course, if a seeded batch crystallization process is to be optimized.

- seeding should be carried out at moderate supersaturation not exceeding 30–40% of the MSZ width;
- fines should always be removed from the seed crystals.

Once the target product property has been defined, the 20 rules given above classify the systems, and as a result, the process conditions, according to the following properties of the system:

- the width of the MSZ;
- the growth rate of the crystals;
- the propensity of the crystals toward attrition; and
- agglomeration.

The rules are divided into three separate decision trees according to the target property of the product crystals. This allows a leaner decision-making process with respect to the process conditions required to optimize a seeded batch crystallization.

References

Costa, C.B.B. and Maciel Filho, R. (2005) Mathematical modelling and optimal control strategy development for an acid crystallization process. *Chem. Eng. Process.*, **44**, 727–753.

Derenzo, S., Shimizu, P.A., and Giulietti, M. (1996) On the behavior of adipic acid aqueos solution batch cooling crystallization, in *Crystal Growth of Organic Materials 3*, American Chemical Society Conference Series (eds A.S. Myerson, D.A. Green, and P. Meenan), American Chemical Society, Washington, DC, pp. 145–150, Proceedings, 27th–31st August, Washington.

Funakoshi, K., Takiyama, H., and Matsuoka, M. (2000) Separations–agglomeration kinetics and product purity of sodium chloride crystals in batch crystallization. *J. Chem. Eng. Jpn.*, **33** (2), 267–272.

Heffels, S. and Kind, M. (1999) Seeding Technology: an underestimated critical success factor for crystallization, in *Proceedings of the 14th International Symposium on Industrial Crystallization, Cambridge, 12th–16th September* (eds J. Garside, R.J. Davey, and M. Hounslow), IChemE Rugby, Cambridge.

Hu, Q., Rohani, S., Wang, D.X., and Jutan, A. (2005) Optimal control of batch cooling seeded crystallizer. *Powder Technol.*, **156** (2–3), 170–176.

Jones, A.G. (1974) Optimal operation of a batch cooling crystallizer. *Chem. Eng. Sci.*, **29**, 1075–1087.

Kind, M. (2004) Grundlagen der technischen kristallisation, in *Kristallisation in der Industriellen Praxis* (ed. G. Hofman), Wiley-VCH Verlag GmbH, Weinheim, pp. 101–112.

Kubota, N. (2001) Seeding policy in batch cooling crystallization. *Powder Technol.*, **121** (1), 31–38.

Kubota, N., Doki, N., Ito, M., Sasaki, S., and Yokota, M. (2002) Seeded batch multistage natural cooling crystallization of potassium alum. *J. Chem. Eng. Jpn.*, **35** (11), 1078–1082.

Miller, S.M. and Rawlings, J.B. (1994) Model identification and control strategies for batch cooling crystallizers. *AIChE J.*, **40** (8), 1312–1327.

Mohamed, H.A., Abu-Jdayil, B., and Al Khateeb, M. (2002) Effect of cooling rate on unseeded batch crystallization of KCl. *Chem. Eng. Process.*, **41**, 297–302.

Mullin, J.W. (1993) *Crystallization*, 3rd edn, Butterworths, London.

Mullin, J.W. and Nyvlt, J. (1971) Programmed cooling of batch crystallizers. *Chem. Eng. Sci.*, **26**, 369–377.

Nyvlt, J. (1988) Programmed cooling of batch crystallizers. *Chem. Eng. Process.*, **24**, 217–220.

Ulrich, M. (1979) Optimierung einer diskontinuierlichen Lösungskristallisation. *Chem. Ing. Tech.*, **51** (3), 243.

Ulrich, J. and Glade, H. (1999) Perspectives in control of industrial crystallizers, *International Workshop "Modelling and Control of Industrial Crystallization Processes" Proceedings*, IBM Centre, La Hulpe.

Warstat, A. (2006) Heuristische regeln zur optimierung von batch-kühlungskristallisationsprozessen. PhD thesis. Martin Luther University Halle-Wittenberg, *http://sundoc.bibliothek.uni-halle.de/diss-online/06/07H060/index.htm*. Accessed 2006.

Warstat, A. and Ulrich, J. (2003) Improvement of product quality using a seeding technique, in *WASIC 2003, Workshop on Advance in Sensoring in Industrial Crystallization* (eds N. Bulutcu, H. Gürbüz, and J. Ulrich), Istanbul Technical University, Istanbul, pp. 34–41.

Warstat, A. and J. Ulrich (2006) Optimierung von Batch- Kühlungskristallisationen, in *Produktgestaltung in der Partikeltechnologie*, vol. **3** (ed. U. Teipel), Fraunhofer IRB Verlag, Stuttgart, pp 413–431.

Warstat, A. and Ulrich, J. (2007) Optimierung von batch-kühlungskristallisationen. *Chem. Ing. Tech.*, **79** (3), 272–280.

Zhang, G.G.Z. and Grant, D.J.W. (2005) Formation of liquid inclusions in adipic acid crystals during recrystallization from aqueous solutions. *Cryst. Growth Des.*, **5** (1), 319–324.

13
Advanced Recipe Control
Alex N. Kalbasenka, Adrie E.M. Huesman, and Herman J.M. Kramer

13.1
Introduction

In advanced recipe control, the heuristic rules used in the basic recipe control are reinforced with the knowledge on the process behavior based on simulations with mathematical models. Process models are used to devise a control strategy that satisfies desired requirements and constraints. This often implies solving an off-line optimization problem with the subsequent open-loop implementation of the obtained optimal input trajectories.

13.2
Incentives and Strategy of the Advanced Recipe Control

It is desired to operate the crystallization process in such a way that the supersaturation stays within the metastable zone. More specifically, crystal growth dominant operation is preferred over nucleation dominant operation as nucleation adversely affects the product quality. Since the goal is to obtain large crystals in many cases, it is beneficial to operate close to the solubility line to minimize the nucleation rate due to the low supersaturation. However, operation under these conditions leads to unrealistically long batch times to obtain crystals of the desired size and reasonable crystal yield. On the other hand, operation close to the metastability limit will lead to an increased nucleation rate. Therefore, an optimization problem can be formulated that seeks a trade-off between a desired fast growth rate and a low nucleation rate.

If the nucleation and the crystal growth rates are both functions of the supersaturation only, then the optimal control strategy amounts to maintaining the supersaturation at a constant level. An operation at a constant supersaturation is often realized for processes dominated by the crystal growth (Hanaki *et al.*, 2007). In more complex cases in which the relation among crystallization phenomena is less obvious, a numerical optimization is needed to design an optimal recipe.

In order to develop an advanced recipe control, the optimal input trajectories can be determined by an off-line optimization using a crystallization model. Although in many cases the incentive for the implementation of a control strategy is improvement of the efficiency of the downstream processes, the optimization objective itself is mostly composed of parameters related to the product quality. The optimal problems are often formulated as maximization of the crystal yield of a batch with a fixed batch time or, alternatively, as minimization of the batch time; maximization of the mean crystal size, while minimizing the width of the crystal size distribution (CSD). It can also be formulated as maximization of a ratio between the crystal volume resulted from the growth of the seed crystals and the volume of crystals grown from generated nuclei.

Seeding parameters (seed mass, size, and the CSD of the seed crystals) are sometimes included as decision variables in a formulation of the optimal problem. It was shown that optimal (nonconservative) seeding followed by an optimal supersaturation control gives better results than a separate application of conservative seeding and supersaturation control. Although the conclusions of those simulation studies are confirmed by various experimental studies (Chianese, Di Cave, and Mazzarotta, 1984), a straightforward implementation of the calculated optimal seeding policies does not always lead to the expected results. In reality, the results can be either better or worse depending on the particular situation. Disagreements between the model-based predictions and the dedicated validation experiments can be due to some practical aspects associated with the preparation of the seed crystals (Kalbasenka *et al.*, 2007); seed characteristics that are generally not accounted for, but have a pronounced impact on the seed growth behavior (origin, state of the crystal surface, and strain content of the crystal lattice); or nucleation mechanisms that became dominant in seeded operation, but were not included in the model formulation (Lewiner *et al.*, 2001).

One of the advantages of using dynamic optimization for designing an optimal recipe is that multiple control objectives can be considered simultaneously. Another benefit is that constraints on the process and control variables and equipment limits can be handled explicitly in the formulation of the optimization problem. Multiple-input multiple-output systems with interacting control loops can also be optimized. Using a mathematical representation of the crystallization process can save time that would be otherwise needed to find the operating policy experimentally.

A mathematical model is the strength and at the same time weakness of model-based control policies. The model should be of a sufficient quality for the optimal results to be beneficial while applied in the reality. The advanced control strategy therefore requires a model that is capable of describing the process dynamics in the given operational envelop.

13.3
Modeling for Optimization, Prediction, and Control

Ideally, it would be beneficial to use one process model for different purposes. This would improve the consistency in the models and minimize the effort needed for the model development and maintenance. However, the model structure depends on the model application and the control objectives that have to be achieved with the help of the model in the particular application (Roffel and Betlem, 2006). Among possible applications of the model are research and development, design, optimization, prediction, and control. It is worth noting that a process model designed for one application is not necessarily suited for another. For example, an excellent model created to support design of a process generally might not be useful for a model-based control of the same process. A transformation would be needed to convert a design model into a model suitable for control. In some cases such as the off-line development of an optimal control strategy, rigorous design models might be also needed.

Since the model is used as a tool assisting in reaching the goal, the model should be fit for the purpose. That is, the model has to be an accurate representation of the modeled system. For optimization, prediction, and control, the model should be capable of accurately predicting the process behavior within the relevant operational envelop.

The choice of model characteristics depends on the control objectives, the targeted application, available knowledge and resources needed for the model development. The following model characteristics need to be considered before modeling the process for dynamic optimization and control purposes:

1) **First-principles modeling versus empirical modeling**.
   The first-principles modeling is based on the conservation laws. It is common practice to use first-principles models based on material and energy balance equations for modeling crystallization processes. Nucleation and growth kinetics are usually modeled using empirical relationships for crystal growth and nucleation. The mechanistic models have an advantage of preserving the physical insight into the process as the model variables coincide with the process variables. This property is very useful while interpreting results of optimization studies. Optimization studies may give a new insight into the process behavior and therefore, may lead to better control policies. A major drawback of physical modeling is that it is a time-demanding modeling approach.
   In empirical modeling, an input–output relationship is constructed using experimental data of the process. Empirical modeling is often used in the circumstances when the time for the model development is limited and/or there is lack of process knowledge. A disadvantage of this approach is that it may require extensive experimentation to obtain the input–output data of sufficient quality for a newly developed process or a process under development.
2) **Distributed-parameter models versus lumped-parameter models**.
   The process model should be as simple as possible. Hence, lumped-parameter models are preferred over distributed-parameter models as they are simpler

and computationally less demanding. Therefore, the most of the optimization studies use a lumped-parameter model and have an objective function formulated from a combination of lumped parameters such as the mean crystal size or the third moment of the CSD. However, if dynamic optimization of the entire CSD is an objective of a model-based control application, a distributed-parameter model with crystal size as a distributed parameter is needed (Worlitschek and Mazzotti, 2004).

Spatial variations in the process variables can be taken into account to build models with descriptive capabilities invaluable for design purposes. Such an approach is often used in the model-based design of crystallizers, but it is not suitable for online repetitive optimization, prediction, and control as the models are too complex to treat them efficiently numerically. The crystallization processes are often considered as well-mixed and therefore modeled as lumped-parameter systems. An assumption on well-mixedness of the crystallizer content is often realistic as uniformity of the crystal slurry is one of the operational goals as described in Section 13.6.

3) **Linear models versus nonlinear models**.
Batch crystallization is a highly nonlinear process. Approximation of the process behavior of batch crystallization over a broad range of operating conditions by linear models can be inaccurate. In the exceptional cases, when the relation between the process input and output is linear or weakly nonlinear, a linear model can be used in a model-based control policy. A necessity of having a linear model may be also dictated by a chosen control algorithm. An example for the application of a linear model in model predictive control (MPC) of fed-batch crystallization is given in Section 14.3.

4) **Dynamic models versus steady-state models**.
Since batch processes are inherently nonstationary, a dynamic model has to be build to obtain model predictions under changing process conditions.

5) **Discrete-time models versus continuous-time models**.
In advanced model-based control configuration, a feedback is introduced to account for model inaccuracies, process disturbances, uncertain initial conditions, and measurement errors. A natural way of representing the process behavior at discrete sampling intervals is to use a discrete formulation of the process model. Since the model has to be implemented on a digital computer, it has to be discretized to compute a numerical solution.

6) **Model complexity**.
Depending on the way in which the model is used in a model-based control system, there can be different requirements on the model complexity. Very detailed models can be effectively utilized for off-line process optimization to obtain an open-loop optimal input trajectory (see Section 13.6). However, if online repeated optimization is targeted, the model should be fast enough to yield an optimal solution within a certain time interval. It is desirable to anticipate the impact that the chosen model structure would have on the computational effort needed to obtain an optimal solution. If the model still

turns out to be too slow, time-consuming model reduction step must be used to reduce the model size.

7) **Model accuracy**.
A sufficiently high accuracy of the model is a prerequisite for a successful implementation of optimal control. Accurate process models are particularly important for open-loop optimal control policies as a large plant-model mismatch can annihilate the benefits of optimal control.

The above-mentioned model characteristics can be contradictory to each other. For example, a good prediction of the model for prediction and control purposes requires a comprehensive description of the process phenomena. A detailed description can lead to a complex and slow model whose usage in online applications for control is problematic. The process characteristics have to be traded off to obtain a balanced model that satisfies the requirements.

13.4
Model Validation

It is assumed at this stage that the process behavior can be qualitatively described by the model. If that is not the case, the underlying model assumptions and the model structure should be verified and adjusted accordingly. A quantitative match of process and model behavior is obtained through the model validation.

One of the sources of uncertainty in the crystallization models are kinetic relations with unknown nucleation and growth constants. In order to ensure that the model is a good representation of the modeled process, the parameters in question need to be estimated from the available experimental data. Batch process data is transient and, in principle, can be used directly for parameter estimation (Tavare, 1987). However, as it was demonstrated in many research studies, a heuristic or model-based (Chung, Ma, and Braatz, 2000) design of dedicated experiments is needed to obtain better estimates for kinetic parameters of empirical (Farrell and Tsai, 1994) or mechanistic (Kalbasenka *et al.*, 2006) kinetic models.

In addition to the experiment design and considerations on the measurements to be used in the parameter estimation, formulation of the estimation problem itself requires particular attention. As parameter estimation is often formulated as an optimization problem, the choice of an objective function and a solution strategy is critical (Beck and Arnold, 1977; Englezos and Kalogerakis, 2001). Furthermore, evaluation of the estimation results is frequently needed to select the most accurate set of the model parameters (Kalbasenka, 2009).

Availability and accuracy of measurements of the solute concentration and the CSD also determine the accuracy of the estimated kinetic parameters. As it was demonstrated by Bermingham (2003), accuracy of the estimates increases when additional information (larger number of the CSD quantiles) on the CSD transients is included in the parameter estimation problem. Moreover, both the measurement of the solute concentration and transients of the CSD are needed for an improved estimation.

13.5
Rate of Supersaturation Generation

Supersaturation is the driving force of crystallization. It influences crystal growth and determines the productivity of batch processes. While it is generally desired to maximize the process yield within a certain batch time, supersaturation is kept at a relatively low level within the metastable zone. The operation above the metastable limit leads to uncontrolled nucleation and, as a consequence, a poor product quality. Therefore, in order to optimize the product quality, an optimal profile is sought for an actuator influencing the rate of supersaturation generation.

In cooling and evaporative crystallization, rate of energy transfer (cooling rate and heating rate needed for evaporation of solvent) has a direct influence on the supersaturation profile. As it was shown by many researchers (see, for instance, Mullin and Nývlt, 1971 and Jones and Mullin 1974), optimal cooling profiles are superior to natural and linear cooling policies as they reduce or avoid the initial peak in the supersaturation trajectory whereby the product quality improves significantly.

In cases, when the process model is not available, the so-called controlled profiles can be used as an approximation of optimal profiles. The controlled cooling profiles were derived using simple crystallization models with the simplified assumptions of negligible nucleation, constant supersaturation, and temperature- and size-independent crystal growth rate (Mullin, 2001).

$$T(t) = T_0 - (T_0 - T_f)\left(\frac{t}{t_f}\right)^a \tag{13.1}$$

in which $a = 3$ for seeded and $a = 4$ for unseeded cooling crystallization.

A relationship similar to Equation (13.1) was proposed for unseeded antisolvent crystallization under assumptions of linear solubility relation and composition-independent nucleation (Tavare, 1995).

$$C(t) = C_0 - (C_f - C_0)\left(\frac{t}{t_f}\right)^4 \tag{13.2}$$

in which C is the antisolvent concentration.

In reactive crystallization, the rate of reactant addition can be found following the same reasoning used for deriving Equations (13.1) and (13.2). Kim et al. (2006) reported the following cubic equation for reactant addition during a seeded reactive crystallization.

$$V(t) = V_{\text{total}}\left(\frac{t}{t_f}\right)^3 \tag{13.3}$$

in which V_{total} is the total volume of reactant that has to be added within the batch time t_f.

The controlled profiles (Equations (13.1)–(13.3)) cannot be applied onto the processes for which the aforementioned simplified assumptions do not apply. In those cases, numerical optimization can be used to find an optimal rate of supersaturation generation as a trade-off between nucleation rate and crystal growth rate.

13.6
Mixing Conditions

Mixing plays an important role in batch crystallization. A poor mixing could be a cause of large temperature and concentration gradients, different local suspension densities, and different CSDs at various locations in the crystallizer. Nonuniformity of the crystal slurry properties could lead to different local nucleation and growth rates that significantly influence the overall performance of a batch crystallization process. As the mixing of the crystallizer content is mostly realized by agitation with an impeller, the impeller rotational speed defines the degree of mixedness, the suspension density distribution, and eventually, the CSD of the product crystals (Sha et al., 1998).

If the crystals are sensitive to contact nucleation, the impeller frequency can also be a decisive factor defining the secondary nucleation rate. Hence, on the one hand, an adequate mixing is required for obtaining a complete suspension of crystals, an efficient energy and mass transfer rates, and a uniform distribution of the suspension density. On the other hand, an intensive mixing is undesirable as it may cause an excessive secondary nucleation yielding product with smaller median sizes and broader distributions. Therefore, an optimization of the stirring policy is necessary to find a compromise between the mixing intensity and the contact nucleation rate. The simulations with optimal impeller frequency profiles show that the product crystals are larger and the CSD is narrower than those resulting from conservative stirring policies (Kalbasenka, Huesman, and Kramer, 2004). As it was shown in the mentioned work, the impact of stirring on the CSD during a batch can be pronounced for systems with relatively heavy and brittle crystals.

An optimal impeller frequency profile can be calculated using a crystallization model in which the nucleation kinetics is accurately modeled as a function of the impeller frequency. In the example given below, a mechanistic secondary-nucleation model of Gahn and Mersmann (1999a, b) was used. An elaborate description of the overall crystallization model is given in the work of Bermingham (2003). The optimal stirring policy was found by solving an optimization problem that maximized the terminal median crystal size on the fixed time horizon of 180 min.

In order to quantify the quality of mixing, some criteria must be defined. Mersmann et al. (1998) introduced two criteria that allow calculating the minimal stirrer speed, which is necessary for the suspension of particles and the local volumetric hold-up of solids. These criteria are the mean specific power input $\bar{\varepsilon}_{BL}$ necessary for off-bottom lifting of the particles and the mean specific power input $\bar{\varepsilon}_{AS}$ necessary to avoid particle settling. Since $\bar{\varepsilon}_{BL}$ is dominant in small vessels and $\bar{\varepsilon}_{AS}$ is the decisive parameter in very large vessels, a general relationship, which combines both criteria, should be used for the vessels of intermediate sizes (as the one we have at hand)

$$\bar{\varepsilon}_{Total} = \bar{\varepsilon}_{BL} + \bar{\varepsilon}_{AS} \tag{13.4}$$

The specific power inputs $\bar{\varepsilon}_{BL}$ and $\bar{\varepsilon}_{AS}$ are functions of the suspension properties (Ar and φ) and crystallizer geometry. The approximate expressions for their calculation are (Mersmann, 2001):

$$\bar{\varepsilon}_{BL} \approx 200 Ar^{1/2} [\varphi(1-\varphi)^n]^{3/4} \frac{v_L g(\rho_C - \rho_L)}{H \rho_L} \left(\frac{T}{D_{imp}}\right)^{5/2} \tag{13.5}$$

$$\bar{\varepsilon}_{AS} \approx 0.4 Ar^{1/8} [\varphi(1-\varphi)^n] \sqrt{d_p \left(\frac{g(\rho_C - \rho_L)}{\rho_L}\right)^3} \text{ with } d_p = 2 \cdot L_{50} \tag{13.6}$$

Comparing the required total specific power input with the power input provided by the impeller provides us with the following requirement to be fulfilled:

$$\bar{\varepsilon} - \bar{\varepsilon}_{Total} > 0 \tag{13.7}$$

where impeller power input is calculated using

$$\bar{\varepsilon} = \frac{4 Ne}{\pi^4} \left(\frac{D_{imp}}{T}\right)^2 \left(\frac{T}{H}\right) \frac{U_{tip}^3}{T} \tag{13.8}$$

If this inequality is satisfied, the first suspension property is met. However, to guarantee a sufficient slurry circulation rate and homogeneity of the suspension, yet another criterion has to be evaluated

$$v_{ax} - 5 U_{slip} > 0 \tag{13.9}$$

in which the axial velocity in the draft tube is calculated using $v_{ax} = 4\phi_{V,DT}/\pi D_{DT}^2$. The concept of particle slip velocity U_{slip} is well known and defined in the work of Zimmels (1983).

Three stirring policies of seeded batch evaporative crystallization of potassium nitrate and ammonium sulfate were studied (Figure 13.5). The initial supersaturation, crystal fraction, and the CSD of the seed population were fixed at the values estimated or measured in batch $DT_c 31$ (see Section 13.8). The kinetic parameters used in optimization cases with $(NH_4)_2 SO_4$ are those estimated for batch $DT_c 31$ (Kalbasenka et al., 2006). In the cases with KNO_3, the kinetic parameters were the same as in the original work of Kalbasenka, Huesman, and Kramer (2004).

The first case simulated the process conducted at a constant impeller frequency of 577 rpm, which was a traditional conservative stirring policy. Two additional stirring policies were obtained by maximizing the terminal median crystal size subject to the rigorous crystallizer model and constraints (Equations (13.7) and (13.9)). In case **a**, a constant optimal frequency was sought, while in case **b**, a time-varying profile parameterized by a piecewise constant function with 10 nonequidistant intervals was obtained. The length of the intervals was an additional optimization variable.

The optimization results for cases with constant and time-varying optimal profiles are shown in Figures 13.3 and 13.4. The corresponding median crystal sizes as

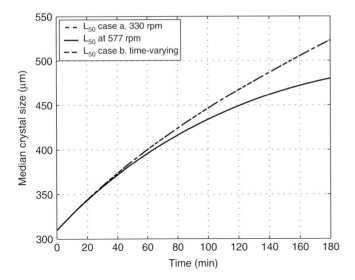

Figure 13.1 Evolution of L_{50} in cases of crystallization of $(NH_4)_2SO_4$.

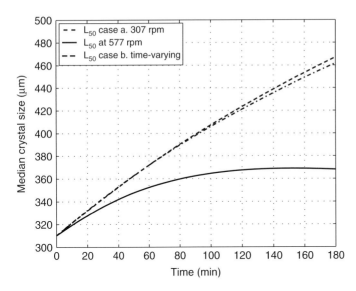

Figure 13.2 Evolution of L_{50} in cases of crystallization of KNO_3.

a function of time are plotted in Figures 13.1 and 13.2. Numerical values of the median crystal size at the process end are summarized in Table 13.1.

The results are consistent with the ones presented in Kalbasenka, Huesman, and Kramer (2004) for the initial conditions corresponding to the first CSD measurement after a nucleation event. The system with relatively hard crystals having a smaller difference in density between the crystals and the solution (ammonium sulfate crystals in its aqueous solution) admit higher stirring rates

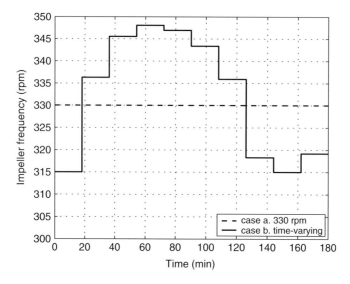

Figure 13.3 Optimal impeller frequency profiles for $(NH_4)_2SO_4$ cases.

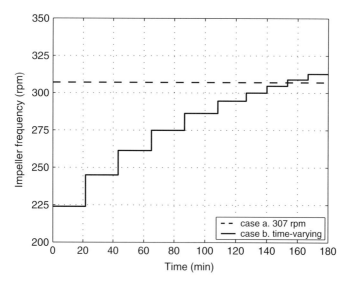

Figure 13.4 Optimal impeller frequency profiles for KNO_3 cases.

in the first half of a batch. Higher stirring rates improve the mass transfer and therefore, promote the crystal growth. At the later stages, a reduction in stirring intensity would be more beneficial as high stirring rates are responsible for an increase in secondary nucleation. The optimal profile is a result of an unconstrained optimization. Because of the flatness of the optimal path, the constant optimal frequency gives the same results as the time-varying one (cf. L_{50} of cases **a** and **b** in Figure 13.1 and Table 13.1).

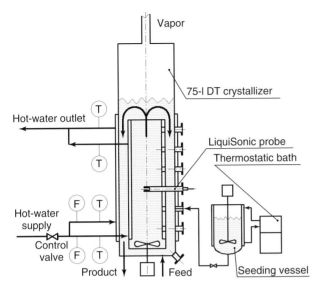

Figure 13.5 75-l DT fed-batch evaporative crystallizer.

Table 13.1 Terminal median crystal sizes at different stirring policies.

Policy	(NH$_4$)$_2$SO$_4$	KNO$_3$
Constant frequency of 577 rpm	480 μm	368 μm
Case a. Optimal constant frequency	523 μm (330 rpm)	462 μm (307 rpm)
Case b. Optimal time-varying policy (Figures 13.3 and 13.4)	523 μm	467 μm

In case of potassium nitrate, the crystals are much more heavy and brittle. The optimal solution follows partly (in case *a*) or completely (in case *b*) the constraint on the velocity difference (Equation (13.9)). This indicates that the process is dominated by the secondary nucleation at all times. The time-varying optimal profile turned out to be slightly better than the constant optimal one (Figure 13.2 and Table 13.1). The difference is not that pronounced as in the cases simulating primary nucleation due to a better management of supersaturation by means of the optimal seeding (Kalbasenka, 2009).

In general, the initial supersaturation, mass, and size distribution of crystals have a large impact on the optimal stirring policy. Hence, a high sensitivity of the optimal results to the initial conditions makes the results of an open-loop implementation of the optimal policies strongly dependent on the reproducibility of the start-up behavior.

As the slurry suspension properties are sensitive to sudden changes in the impeller frequency, it is recommended to implement the open-loop optimal profiles using a detuned proportional-integral (PI) controller to realize a smooth impeller-frequency trajectory. Application of seeding can improve reproducibly of batch operation and control of the initial supersaturation profile (Chapter 11) and therefore, a constant optimal stirring rate can be applied without a loss in the process performance characteristics.

13.7
Implementation

The optimal profiles can be implemented using sequential tables containing time profiles of the optimized process input. The values of the optimal trajectory are passed as a set-point onto a servocontroller, which is usually a proportional-integral-derivative (PID) controller.

Fujiwara *et al.* (2005) suggest that instead of using time profiles for generating the set-point for a PID controller, the set-point calculation should be done using optimal relationship between the process input and the influenced process variable. In batch cooling crystallization, for example, a time-dependent optimal trajectory of the crystallizer temperature should be replaced by an equivalent concentration–temperature trajectory. Such implementation is thought to be robust to process disturbances as it takes into account the state of the system. The drawback of this strategy is that the measurement of the process variable (such as concentration) should be available during the batch. While in industries such as sugar-producing industry, a reliable *in-situ* concentration measurement is common practice, measuring concentration in other industries is problematic due to, for instance, a low accuracy or high costs.

13.8
Example of Modeling, Optimization, and Open-Loop Control of a 75-l Draft-Tube Crystallizer

13.8.1
Objectives and Advanced Recipe Control

The main control objective of this study was to maximize the process yield by manipulating the rate of energy transfer. The product quality expressed in terms of the median crystal size and the CSD width had to be similar or better than in the process with a constant energy-transfer rate. Next to the product quality-related constraints, hardware-related constraints had to be satisfied.

The advanced recipe consisted of two parts – the batch initiation by seeding and the process operating according to an open-loop optimal supersaturation generation. A nonconservative seeding procedure was developed using the guidelines given in Chapter 11 and experience. The adopted seeding procedure enabled control over the initial population of crystals and the desupersaturation profile. In order to guarantee the survival of the seed crystals, a lower limit on the energy-transfer rate and the initial supersaturation were determined.

Given the defined initial conditions, the operational and quality constraints, and the control objectives, an objective function was formulated. Using a process model, an optimization problem was formulated and solved to yield an optimal input profile. The optimal trajectory was implemented in an open-loop fashion onto the actual crystallization process.

13.8.2
Process Description and Modeling

A 75-l DT fed-batch evaporative crystallizer was modeled as a single well-mixed compartment with one-inlet and two-outlet streams (Figure 13.5). The outlet streams were an unclassified product removal stream F_p and a crystal- and solute-free vapor stream. A small product removal was required for online measurement of the CSD with a laser diffraction instrument (HELOS-Vario, Sympatec, Germany). The single-inlet flow was the crystal-free feed stream containing ammonium sulfate solution saturated at 50 °C. The feed stream was used to keep the crystallizer volume constant by compensating for losses in volume due to evaporation of the solvent and the slurry sampling. The crystallizer was operated isothermally at 50 °C. The material properties were calculated for this temperature. They were assumed constant during the batch time. The crystallizer impeller was kept at a constant frequency of 450 rpm.

The crystallizer was seeded with 600 g of seeds. The seeds were obtained by milling and sieving commercial product crystals of ammonium sulfate (DSM, the Netherlands). The necessary amount of the sieve fraction of 90–120 μm was collected. The seed crystals were aged for an hour in a saturated solution of ammonium sulfate in the seeding vessel (Kalbasenka et al., 2007). The seeds were introduced at a predetermined supersaturation level that was measured by an ultrasonic analyzer (LiquiSonic®, SensoTech, Germany).

A population balance equation (PBE) for a fed-batch evaporative, isothermal, well-mixed crystallizer with an unclassified product removal (for sampling purposes only) and a crystal-free feed reads

$$\frac{\partial n(t, L)}{\partial t} = -G \frac{\partial n(t, L)}{\partial L} - \frac{F_p}{V} n(t, L) \tag{13.10}$$

with the following boundary conditions:

$$n(0, L) = n_0(L) \tag{13.11}$$

$$n(t, 0) = \frac{B_0}{G} \tag{13.12}$$

with $n_0(L)$ being the CSD of the seed crystals at the moment of seeding (not to be confused with the CSD of the sieve fraction of the seeds as it can undergo considerable changes during the seed preparation (Kalbasenka et al., 2007)) and B_0 is the nucleation rate in which index 0 refers to the birth of infinitesimal crystals (with size $L = 0$).

Due to the dominance of crystal growth over secondary nucleation in batch runs with a large seed load, secondary nucleation by crystal attrition was neglected. The crystal growth rate was assumed to be independent of crystal size. As a consequence thereof, the PBE could be reduced to a set of ordinary differential equations by multiplying both sides of Equation (13.10) by $L^i dL$, integrating from zero to infinity, and taking the first five moments of the CSD (Randolph and Larson, 1971).

The advantage of using the moment model is its simplicity. It does not require much computation time and therefore, does not need further model reduction. The latter property of the moment model is crucial for online dynamic optimization of the process performance as described in Section 14.2.

After the transformation of the PBE as described above, the moment model of the fed-batch process takes the form of the following expressions involving the first five moments of the CSD:

$$\frac{dm_i}{dt} = 0^i \cdot B_0 + iGm_{i-1} - \frac{m_i F_p}{V} \quad \text{with } i = 0, \ldots, 4 \quad (13.13)$$

with the initial conditions given by

$$m_i(0) = m_{i,0}, \quad i = 0, \ldots, 4 \quad (13.14)$$

Using mass and enthalpy balance equations, one can derive a single expression for the solute concentration. In this case, it is of the following form:

$$\frac{dC}{dt} = \frac{F_p(C^* - C)/V + 3k_v G m_2 (k_1 + C)}{1 - k_v m_3} + \frac{k_2 Q}{1 - k_v m_3} \quad (13.15)$$

with the constant coefficients k_1 and k_2 given by

$$k_1 = \frac{H_v C^*}{H_v - H_L} \left(\frac{\rho_c}{\rho_L} - 1 + \frac{\rho_L H_L - \rho_c H_c}{\rho_L H_v} \right) - \frac{\rho_c}{\rho_L}$$

$$\text{and} \quad k_2 = \frac{C^*}{V \rho_L (H_v - H_L)} \quad (13.16)$$

and the initial condition defined by

$$C(0) = C_0 \quad (13.17)$$

The nucleation rate B_0 and the size-independent crystal growth rate G are expressed using simple empirical relationships:

$$B_0 = k_N m_3 G \quad (13.18)$$

with

$$G = k_G \Delta C = k_G (C - C^*) \quad (13.19)$$

Kinetic parameters k_N and k_G needed to be estimated using dynamics experimental data of the batch process (experiment $DT_c 31$). The used experimental data included information on the crystal fraction and the CSD. The evolution of crystal fraction was inferred from the density measurements ρ_p of the unclassified product stream assuming that the density of liquid phase of the crystal slurry was equal to the

13.8 Example of Modeling, Optimization, and Open-Loop Control of a 75-l Draft-Tube Crystallizer

Table 13.2 Numerical values of variables and parameters.

Symbol	Variable or parameter	(Initial) Value	Unit
B_0	Nucleation rate	Use Equation (13.18)	#/(m^3 s)
C	Solute concentration	0.457170	kg solute/kg solution
C^*	Saturation concentration at 50 °C	0.456513	kg solute/kg solution
G	Crystal growth rate	Use Equation (13.19)	m/s
$G_{\max}$	Maximum growth rate	2.5×10^{-8}	m/s
H_c	Specific enthalpy of crystals	69.8625	kJ/kg
H_L	Specific enthalpy of liquid	60.7500	kJ/kg
Q	Heat input	9.0 (in batch DT$_c$31)	kW
H_v	Specific enthalpy of vapor	2.590793×10^3	kJ/kg
k_v	Volumetric shape factor	0.43	–
k_N	Nucleation rate constant	1.02329×10^{14}	#/m^4
k_G	Growth rate constant	7.49567×10^{-5}	m/s
m_0	Zeroth moment of CSD	1.291148×10^{10}	1/m^3
m_1	First moment of CSD	2.663270×10^4	m/m^3
m_2	Second moment of CSD	3.581689×10^2	m^2/m^3
m_3	Third moment of CSD	8.902700×10^{-2}	m^3/m^3
m_4	Fourth moment of CSD	2.945600×10^{-5}	m^4/m^3
F_p	Product flow rate	1.614950×10^{-6}	m^3/s
V	Crystallizer volume	7.50×10^{-2}	m^3
ρ_c	Density of crystals	1248.9	kg/m^3
ρ_L	Density of saturated solution	1767.3	kg/m^3

density of the saturated solution at the same temperature (Table 13.2). Then, the crystal volume fraction can be calculated as

$$\varepsilon_c = \frac{\rho_p - \rho_L}{\rho_c - \rho_L} \tag{13.20}$$

The CSD was measured by the laser diffraction instrument. The volume-based CSDs were converted to the moments of the CSD. The second, the third, and the fourth moments of DT$_c$31 were used to estimate the nucleation and growth kinetic parameters as the measurement of those moments were more reliable than the signals of the zeroth and the first moments. The initial values of the zeroth and the first moments were estimated together with the kinetic parameters. The initial values of some variables and values of constants are given in Table 13.2.

13.8.3
Dynamic Optimization

The following optimization problem was solved in MATLAB® using function *fmincon*:

$$\min_{Q(t), t \in [0 \ t_f]} J(t_f) \tag{13.21}$$

subject to Equations (13.13)–(13.19) and $9 \leq Q(t) \leq 13$ kW. The heat-input trajectory was parameterized using a piecewise-constant function of time with equidistant time intervals. It is worth noting that in some optimization algorithms, the choice of representation (discretization or parameterization) and the initial estimate of the control vector can have a significant influence on the optimization results (Costa and Filho, 2005). As it is illustrated in Ward, Mellichamp, and Doherty (2006), the choice of the objective function is also critical for obtaining an optimal trajectory that is adequate to the posed control objectives.

Function *fmincon* is a sequential quadratic programming (SQP) optimization algorithm that is most frequently used for optimization of process systems (Edgar, Himmelblau, and Lasdon, 2001). However, applications of other algorithms to optimization of crystallization processes were reported in the literature (the steepest descent method (Shi et al., 2006) and a genetic algorithm (Costa and Filho, 2005)). The choice of optimization method is important for obtaining an accurate global optimal solution. In addition, this choice becomes particularly important for online repeated dynamic optimization (Section 14.2).

The lower limit on the heat input was necessary to guarantee a high survival efficiency of the seed crystals by maintaining a relatively high level of supersaturation at the beginning of batch (Kalbasenka et al., 2007). An optimal heat-input trajectory was sought on the time interval $[\ 0 \quad t_f\ ]$ with a fixed batch time ($t_f = 180$ min). The objective function $J(t_f)$ was of the following form:

$$J(t_f) = \frac{W \int_0^{t_f} \left(100 \frac{G(t) - G_{max}}{G_{max}}\right)^2 dt}{\int_0^{t_f} dt} \quad \left\{ \begin{array}{lll} \forall t: & G(t) - G_{max} < 0; & W = 1 \\ \forall t: & G(t) - G_{max} \geq 0; & W = 10 \end{array} \right\} \quad (13.22)$$

The crystal growth rate was calculated using Equation (13.19). The objective function $J(t)$ was designed to enable operation close to the maximum growth rate G_{max}. Because of the maximization of the crystal growth, the process yield could be increased as well. Such a formulation of the objective function also allowed for maximizing the median crystal size since higher crystal growth rates were strived for. An upper constraint on the crystal growth rate was imposed in order to limit nucleation. The maximum growth rate was fixed at the value of $G_{max} = 2.5 \times 10^{-8}$ m/s, above which higher nucleation rates were observed experimentally. As the crystal growth rate is a linear function of supersaturation, a constant supersaturation control was in fact sought for in the region where limits on the heat input were inactive. The optimal solution is then a trade-off between the nucleation rate and the crystal growth rate.

The objective was to keep the process at the constraint if feasible. G_{max} was formulated as the so-called soft constraint. That is, short excursions above that limit are undesired, but allowed. The weights W in the objective function (Equation (13.22)) were chosen in such a way that a penalty for excursions above this constraint was larger than for deviations below it. The overshoots were suppressed more heavily ($W = 10$) than undershoots ($W = 1$) to prevent an excessive nucleation from happening.

13.8.4
Experimental Validation Results

The optimal profile obtained by off-line optimization was implemented in an open-loop mode as a time-varying set-point of the PI heat-input controller. The process value of the heat input was calculated from the measurements of flow rate and inlet and outlet temperatures of the streams supplying hot water to the crystallizer jackets. In order to track the given set-point values, the PI controller was manipulating the control valve to adjust the total flow rate of hot water supply (Figure 13.5).

The experimental results of the open-loop implementation of the optimal trajectory are presented in Figures 13.6–13.9 (experiment DT_c55). The PI controller was capable of the following the given set-point rather well as it can be seen in Figure 13.6.

The growth rate was calculated using supersaturation values computed by a state estimator that is further introduced in Section 14.2. From the calculated values of the growth rate, it can be concluded that higher growth rates were achieved in the second half of batch DT_c55 compared with batch DT_c31 during which the heat input was kept at 9 kW (Figure 13.7).

The variance of the crystal growth rate could have been reduced if an optimal heat-input profile with a finer temporal resolution had been implemented. However, due the open-loop implementation of the optimal trajectory, the process operation close to the constraint could not be guaranteed (see Figure 13.7). As it is shown in Section 14.2, the shortcomings of the open-loop implementation

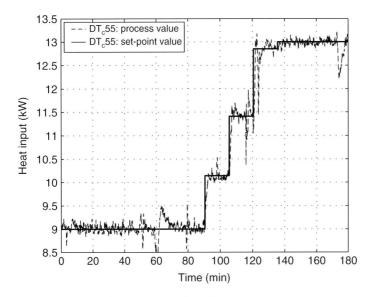

Figure 13.6 Optimal vs. implemented profile.

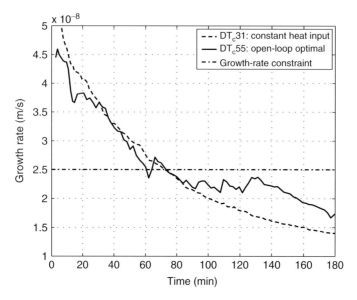

Figure 13.7 Growth rate during batch DT_c55.

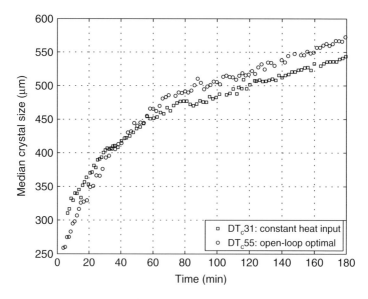

Figure 13.8 Comparison of the median sizes.

in this particular example are due to discrepancies between the initial conditions used in derivation of the open-loop optimal trajectory and actually observed values.

Despite some inaccuracies of the open-loop implementation, the obtained results were superior to those of the batches conducted at a constant heat input. Higher

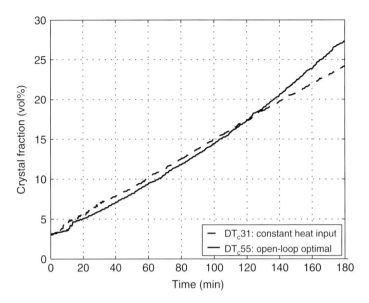

Figure 13.9 Comparison of the process yields.

growth rates achieved in batch DT_c55 led to a 12.8% increase (3.1 vol% of the crystal fraction) in the process yield as it is shown in Figure 13.9. As it was expected, the median crystal size was larger too (Figure 13.8). The width of the CSD did not undergo much change. Therefore, no significant increase in nucleation was induced by the open-loop application of the optimal heat-input trajectory.

13.9 Conclusions

An optimization of the process operation using validated process models allows for derivation of an advanced control recipe. The optimization problem can include a complex objective function to obtain an optimal solution in cases when simplified assumptions are not valid. An optimal solution can also be obtained in the presence of constraints on the process variables and the actuators. Moreover, the model-based optimization studies can increase understanding of the studied process and can save resources needed to arrive at an optimal recipe experimentally. However, the process modeling and optimization can in turn be time- and expertise-demanding activities.

Advantages of an open-loop implementation of the nominal optimal trajectory can be diminished due to a disagreement in the initial conditions, model and measurement errors, and nonmeasurable process disturbances. Using process measurements directly in the formulation of an advanced model-based control policy by introducing feedback can compensate for the shortcomings of the open-loop strategies (Chapter 14).

References

Beck, J. and Arnold, K. (1977) *Parameter Estimation in Engineering and Science*, John Wiley & Sons, Inc., New York.

Bermingham, S. (2003) A design procedure and predictive models for solution crystallisation processes. PhD thesis. Delft University of Technology, The Netherlands, *http://www.library.tudelft.nl* (accessed 2011).

Chianese, A., Di Cave, S., and Mazzarotta, B. (1984) Investigation on some operating factors influencing batch cooling crystallization, in (eds S. Jančić and E. de Jong), *Proceedings of the 9th Symposium on Industrial Crystallization, The Hague, September 25–28*, Elsevier, Amsterdam, pp. 443–446.

Chung, S., Ma, D., and Braatz, R. (2000) Optimal model-based experimental design in batch crystallization. *Chemom. Intell. Lab. Syst.*, **50**, 83–90.

Costa, C. and Filho, R. (2005) Evaluation of optimisation techniques and control variable formulations for a batch cooling crystallization process. *Chem. Eng. Sci.*, **60**, 5312–5322.

Edgar, T., Himmelblau, D., and Lasdon, L. (2001) *Optimization of Chemical Processes*, McGraw-Hill Chemical Engineering Series, 2nd edn, (eds Glandt, E., Klein, M., and Edgar, T.), McGraw-Hill, New York.

Englezos, P. and Kalogerakis, N. (2001) *Applied Parameter Estimation for Chemical Engineers*, Marcel Dekker, New York.

Farrell, R. and Tsai, Y.-C. (1994) Modeling, simulation and kinetic parameter estimation in batch crystallization processes. *AIChE J.*, **40** (4), 586–593.

Fujiwara, M., Nagy, Z., Chew, J., and Braatz, R. (2005) First-principles and direct design approaches for the control of pharmaceutical crystallization. *J. Process Control*, **15**, 493–504.

Gahn, C. and Mersmann, A. (1999a) Brittle fracture in crystallization processes. Part A. Attrition and abrasion of brittle solids. *Chem. Eng. Sci.*, **54**, 1273–1282.

Gahn, C. and Mersmann, A. (1999b) Brittle fracture in crystallization processes. Part B. Growth of fragments and scale-up of suspension crystallizers. *Chem. Eng. Sci.*, **54**, 1283–1292.

Hanaki, K., Nonoyama, N., Yabuki, Y., Kato, Y., and Hirasawa, I. (2007) Design of constant supersaturation cooling crystallization of a pharmaceutical: a simple approach. *J. Chem. Eng. Jpn.*, **40** (1), 63–71.

Jones, A. and Mullin, J. (1974) Programmed cooling crystallization of potassium sulphate solutions. *Chem. Eng. Sci.*, **29**, 105–118.

Kalbasenka, A.N. (2009) Model-based control of industrial batch crystallizers: experiments on enhanced controllability by seeding actuation. PhD thesis. Delft University of Technology, The Netherlands. ISBN: 978-90-6464-361-3, *http://www.dcsc.tudelft.nl/Research/PhDtheses/*.

Kalbasenka, A., Huesman, A., and Kramer, H. (2004) Impeller frequency as a process actuator in suspension crystallization of inorganic salts from aqueous solutions, in *11th International Workshop on Industrial Crystallization, September 15–17, Gyeongju* (ed. K.-J. Kim), Hanbat National University, Daejeon, pp. 135–143.

Kalbasenka, A., Huesman, A., Kramer, H., and Bosgra, O. (2006) On improved experiments for estimation of Gahn's kinetic parameters, in *The 13th International Workshop on Industrial Crystallization, September 13–15, Delft* (eds P. Jansens, J. ter Horst, and S. Jiang), IOS Press, Amsterdam, pp. 122–129.

Kalbasenka, A., Spierings, L., Huesman, A., and Kramer, H. (2007) Application of seeding as a process actuator in a model predictive control framework for fed-batch crystallization of ammonium sulphate. *Part. Part. Syst. Charact.*, **24** (1), 40–48.

Kim, S., Lotz, B., Lindrud, M., Girard, K., Moore, T., Nagarajan, K., Alvarez, M., Lee, T., Nikfar, F., Davidovich, M., Srivastava, S., and Kiang, S. (2006) Control of the particle properties of a drug substance by crystallization engineering and the effect on drug product formulation. AIChE Spring Meeting Conference Proceedings, Vol. 2: Fifth World Congress on Particle Technology, April 23–27, Orlando, FL.

Lewiner, F., Févotte, G., Klein, J.P., and Puel, F. (2001) Improving batch cooling seeded crystallization of an organic weed-killer using on-line ATR FTIR measurement of supersaturation. *J. Cryst. Growth*, **226**, 348–362.

Mersmann, A. (ed.) (2001) *Crystallization Technology Handbook*, 2nd edn, Marcel Dekker, New York.

Mersmann, A., Werner, F., Maurer, S., and Bartosch, K. (1998) Theoretical prediction of the minimum stirrer speed in mechanically agitated suspensions. *Chem. Eng. Process.*, **37**, 503–510.

Mullin, J. (2001) *Crystallization*, 4th edn, Butterworth-Heinemann, Oxford.

Mullin, J.W. and Nývlt, J. (1971) Programmed cooling of batch crystallizers. *Chem. Eng. Sci.*, **26**, 369–377.

Randolph, A. and Larson, M. (1971) *Theory of Particulate Processes*, Academic Press, New York.

Roffel, B. and Betlem, B. (2006) *Process Dynamics and Control: Modeling for Control and Prediction*, John Wiley & Sons, Ltd, Chichester, UK.

Sha, Z., Louhi-Kultanen, M., Ogawa, K., and Palosaari, S. (1998) The effect of mixedness on crystal size distribution in a continuous crystallizer. *J. Chem. Eng. Jpn.*, **31** (1), 55–60.

Shi, D., El-Farra, N., Li, M., Mhaskar, P., and Christofides, P. (2006) Predictive control of particle size distribution in particulate processes. *Chem. Eng. Sci.*, **61**, 268–281.

Tavare, N. (1987) Batch crystallization: a review. *Chem. Eng. Commun.*, **61**, 259–318.

Tavare, N. (1995) *Industrial Crystallization: Process Simulation Analysis and Design*, The Plenum Chemical Engineering Series (ed. D. Luss), Plenum Press, New York.

Ward, J., Mellichamp, A., and Doherty, M. (2006) Choosing an operating policy for seeded batch crystallization. *AIChE J.*, **52** (6), 2046–2054.

Worlitschek, J. and Mazzotti, M. (2004) Model-based optimization of particle size distribution in batch-cooling crystallization of paracetamol. *J. Cryst. Growth*, **4** (5), 891–903.

Zimmels, Y. (1983) Theory of hindered sedimentation of polydisperse mixtures. *AIChE J.*, **29**, 669–676.

14
Advanced Model-Based Recipe Control
Alex N. Kalbasenka, Adrie E.M. Huesman, and Herman J.M. Kramer

14.1
Introduction

In order to cope with shortcomings of the open-loop optimal control due to model inaccuracies and process disturbances, closed-loop optimal control strategies can be used as discussed in the previous chapter. The main characteristic feature in those strategies is that a process model is used as an integral part of the model-based control system. The second most important characteristic of closed-loop control methods is that an interaction between the control system and the process is introduced by means of a real-time measurement feedback.

The process model can be used in a number of ways within the model-based recipe control. The model can be identified from simulated or experimental data and used in a model-based controller as described in Chapter 16. However, the most common way of using the model is to carry out online optimization of the crystallization process using a first-principles nonlinear model with empirical kinetic relations or a linear model derived from the nonlinear one (see Chapter 13 and the examples below).

A few applications of online closed-loop optimal control strategies to batch crystallization are reported in the literature. Most of them are concerned with simulations of the optimal control applied to cooling crystallization.

Shi *et al.* (2006) simulated a closed-loop predictive control strategy with repeated optimization minimizing the volume of fines in the final product in seeded batch cooling crystallization of potassium sulfate. A population balance equation (PBE) based model was used to simulate the process measurements, while a moment model was employed for prediction in the MPC. They showed that maximizing the ratio between the third moments corresponding to crystals grown from seeds and generated by nucleation can indeed lead to the product with larger volume of seed-grown crystals, but the volume of fines increases dramatically at the same time. An alternative optimization strategy that gave better results minimized the third moment of the nucleated crystals.

Zhang and Rohani (2003, 2004) considered online optimal control of a seeded batch crystallizer. A dynamic optimization problem involved a composite objective function that maximized the weight mean crystal size and minimized the coefficient of variation at the same time. In addition, the process states were assumed to be nonmeasurable. An extended Kalman filter was designed to observe the states from the simulated process data. A proportional-integral (PI) controller was included to track the optimal trajectory generated by the dynamic optimizer.

Tracking of an off-line-derived optimal cooling trajectory with servo-controllers was studied by Shen, Chiu, and Wang (1999). They found that using globally linearizing control, generic model control (see also the work of Vega, Diez, and Alvarez (1995) for experimental results), and multimodel-based MPC for the design of a tracking trajectory gives superior results than using a conventional PI controller for the same goal. Because of the direct utilization of the process model in the formulation of the model-based controllers, a better performance was also achieved with respect to disturbance rejection and robustness to the modeling errors.

To summarize, in all those strategies, a process model is used online to compensate for uncertainties in initial conditions, unmeasured process disturbances, process–model mismatch, and noisy or unmeasured process states. Next to the optimization algorithm and the model-based controller formulation, state estimation employs the process model to reconstruct the state variables and unmeasured process variables from the available (noisy) process measurements. A state estimator or an observer is also used to detect unmeasured process disturbances. The up-to-date information about the system states is utilized by an optimizer to re-compute the optimal operating policy online.

The common feature of the closed-loop strategies described above is that they perform an explicit or an implicit measurement-based optimization of the process conditions with respect to some economic or operational criteria (Bonvin, Srinivasan, and Ruppen, 2002). The differences in the two methods are determined by the way the model is used to optimize the process. The first group of control policies uses the process model and measurements for online repeated optimization (explicit optimization of the process performance formulated as an optimization problem with an objective function similar to the ones given by Equations (13.21) and (13.22)) for adaptation of the input trajectory to ensure optimality in the presence of uncertainties.

The control strategies that use a model-based tracking controller to follow an off-line-generated optimal profile constitute the second group. The optimization of the process performance is implicit since the optimal reference trajectory is computed off-line in a similar way as described in Chapter 13. The second group of methods differs from the open-loop optimal control methods of Chapter 13 in the way the tracking of the trajectory is realized. In the former case, a model-based controller uses the process measurements to follow the optimal trajectory. The model employed in the controller needs not be the same as used in the derivation of the optimal path. The tracking controller involves online optimization to compute a control action needed to realize an optimal tracking in the presence of uncertainties. In that case, the optimal tracking problem involves a cost function formulated in terms of deviations of the tracked variables and the controls from their optimal paths. Control methods that use the batch-end measurement to improve the open-loop optimal trajectory and by doing so, minimize batch-to-batch variations fall under the second category.

In the next sections, applications of the both types of closed-loop optimal control policies are described.

14.2
Online Dynamic Optimization

In this section, an application of online repeated dynamic optimization is illustrated for the evaporative crystallization process introduced in Section 13.8. The dynamic optimization problem (Equation (13.21)) was solved online repeatedly in a receding horizon mode (Chapter 16). The process measurements were used to estimate the initial conditions needed to initialize optimizations. An optimal profile was implemented onto the process using a conventional PI controller. A schematic representation of the advanced model-based control strategy is shown in Figure 14.1.

The operation principle of the control system with the dynamic optimizer is similar to the receding horizon principle introduced in Chapter 16. The dynamic optimizer was initialized with the values corresponding to the process conditions at the time of the first CSD measurement. The computed optimal trajectory $\mathbf{Q}_{opt}(k+j)$ consisted of 15 two-minute intervals j spanning 30 min of the prediction horizon. The first element of the trajectory $Q_{opt}(k)$ was used as a set-point value Q_{sp} for a PI controller of the heat input. The process value of the heat input was subtracted from the set-point value to yield an error e. The PI controller calculated a control action u needed to eliminate the error and bring the process value to the set-point. For the process at hand, u was the position of the control valve that regulated the flow of hot water to the crystallizer jackets (see Figure 13.5). The control action was implemented and the entire calculation sequence was repeated after the time of a sample interval was elapsed.

The advanced model-based control architecture was similar to the one presented in the diagram (Figure 16.5) in Section 16.9 (except that the dynamic optimizer was used instead of the MPC controller). The heat-input controller was a part of the DCS system (CENTUM CS3000, Yokogawa, Japan). A timed signal exchange among the modules (the dynamic optimizer, the observer, and the DCS) was enabled by a timer and an OPC server (IPCOS BV, the Netherlands).

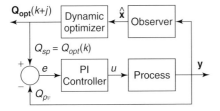

Figure 14.1 Block diagram of the online closed-loop optimal control system.

The prediction horizon on which the optimal input trajectory is computed is a tuning parameter similar to the one in the MPC controller. Generally, the longer the prediction horizon is, the smoother the optimal input trajectory is. However, optimization on long prediction horizons can be computationally expensive. Since the optimal solution has to be obtained within a certain time interval, the length of the prediction horizon can be tuned to find a balance between the computation time and the acceptable controller performance (see Section 16.7 for more details on tuning the controller performance).

The length of the sample interval is also a tuning parameter (Agachi et al., 2006). Its duration is a compromise between the necessity of capturing the fast process dynamics and having sufficient time for computing the optimal input trajectory. Shi et al. (2006) showed that in case of an imperfect model, a tighter feedback control was needed to avoid constraint violation. The tighter feedback control was achieved with a smaller sample interval. Hence, there is a direct relationship between the model accuracy and the maximum sample interval. In the considered application, the sample interval was set equal to the duration of the CSD measurement cycle (120 s).

For the given process, the measured variables were the volume-based distributions measured by HELOS-Vario, the density of the product slurry, and the current process value of the heat input Q_{pv}. The leading moments of the CSD were calculated online from the volumetric size distributions and the volume fraction of crystals. The volumetric crystal fraction was computed from the slurry density using expression (13.20).

A state estimator or an observer (Kalbasenka, 2009) was used to estimate the process state variables including unmeasured solute concentration. The observer was based on the nonlinear moment model presented in Section 13.8 (Equations (13.13)–(13.19)). As the first two CSD moments m_0 and m_1 are sensitive to measurement errors in the volume-based CSD from which they are calculated, the moments m_2, m_3, and m_4 were used by the observer to produce an estimate of the state variables (all the CSD moments and the solute concentration). The estimated states were used for the initialization of the dynamic optimization and computation of an optimal input trajectory by solving the minimization problem (Equation (13.21)) online.

The results of the described closed-loop optimal control strategy with the dynamic optimizer are shown in Figures 14.2 and 14.3. In Figure 14.2, the optimal profile used as a time-varying set-point for the base layer PI controller of the heat input is compared with the input profile that was actually applied to the system. The PI controller was able to follow the desired trajectory rather closely. It is worth mentioning that the optimal set-point profile in Figure 14.2 is composed of the first elements of the optimal trajectories that were generated by the optimizer each time new process measurements became available. In this way, optimality of the process performance was guaranteed even in the presence of uncertainties.

The optimal input profile in Figure 14.2 has three distinct intervals or arcs. The optimal solutions on the first and the third interval are defined by the lower and the upper bound of the heat input, respectively. The lower bound was imposed

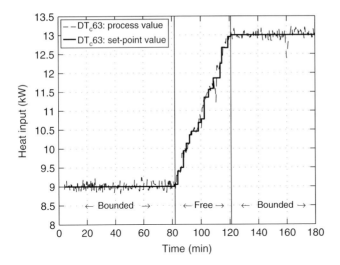

Figure 14.2 Optimal vs. implemented profile.

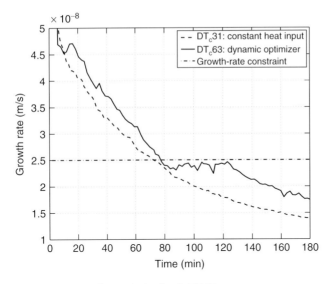

Figure 14.3 Growth rate during batch DT_c63.

to guarantee the survival of seed crystals (Kalbasenka et al., 2007). The upper bound is a hardware constraint. So, the optimal solution of those arcs is bounded (Figure 14.2). The optimal solution of the second interval is free of the input bounds and defined by a path constraint. The path constraint is in fact a state constraint. As it can be seen from an equivalent formulation of the optimization problem (Equation (13.21)), the constrained state is the solute concentration:

$$\max_{Q(t), t \in [0\ t_f]} \varepsilon_c(t_f) \tag{14.1}$$

subject to Equations (13.13)–(13.19), $9 \leq Q(t) \leq 13$ kW, and $G(t) < G_{max}$.

Substitution of Equation (13.19) into the path constraint of the optimization problem (Equation (14.1)) yields

$$C(t) - C_{max} < 0 \qquad (14.2)$$

where $C_{max} = C^* + G_{max}/k_G = 0.456847$ kg solute/kg solution in which C^* is assumed to be constant (see Table 13.2). Therefore, evolution of the solute concentration has a profile similar to that of the crystal growth rate in Figure 14.3.

From the above analysis of the optimal trajectory, it follows that the process should be operated as close as possible to the growth rate constraint in order to ensure the optimality of the operation and maximize the process yield.

The open-loop implementation of the optimal heat-input profile described in Chapter 13 suffered from the process disturbances. A process disturbance occurred at the 60th min of batch DT_c55. Since that disturbance was not compensated for, a large variance in the process variables was induced. As the process was operated close to the state constraint, the large variance of the growth rate signal caused an overshoot above the maximum growth rate (Figure 13.7). Moreover, the growth rate constraint was not followed closely due to uncertainties in the initial conditions. Therefore, the process operation was not optimal. Introduction of state feedback allows for detection and compensation for the process disturbances as well as the uncertainties in the initial conditions. The process operation is then more optimal as the reduced variance in the process variables permits a closer tracking of the growth rate constraint without violating it (cf. Figures 13.7 and 14.3). As a result, the crystal fraction could be increased by an additional 1 vol% (approximately 3.8% of the total process yield) without loss in the product quality (cf. Figures 13.8 and 14.8, see also Table 14.1).

As industrial processes are rarely free of uncertainties, disturbances, and measurement noise, closed-loop control strategies are preferred over open-loop ones.

The advantage of implementing the online optimization of the heat-input profile can be further illustrated by comparing the implemented trajectory with an

Table 14.1 Conditions and experimental results of the process control validation studies.

Batch number	Heat input[a] (kW)	Initial supersaturation at seeding[b] (–)	Terminal crystal fraction (vol%)	Terminal median crystal size (μm)	Terminal width of the CSD[c] (–)
DT_c31	9.0	0.01627	24.2	544	2.45
DT_c55	As in Figure 13.6	0.00567	27.3	573	2.62
DT_c62	As in Figure 14.6	0.00694	28.0	579	2.53
DT_c63	As in Figure 14.2	0.00718	28.3	565	2.56

[a] All experiments were conducted with the impeller frequency kept at 450 rpm.
[b] All batches were seeded with the seed slurry prepared from 600 g of 90–125-μm sieve fraction according to the procedure described in Kalbasenka et al. (2007) and Kalbasenka (2009).
[c] The CSD width is defined as the ratio of the 90% quantile to 10% quantile of the CSD.

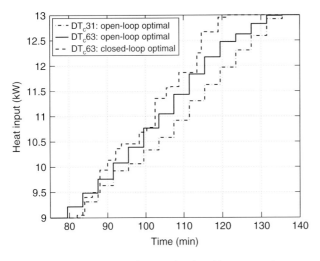

Figure 14.4 Open-loop profiles vs. the closed-loop optimal trajectory of DT_c63.

open-loop optimal one. In Figure 14.4, the optimal trajectory of batch DT_c63 is compared to an open-loop optimal trajectory. The open-loop profile was computed using initial values of the state variables estimated from the first CSD measurement of the same batch. The heat input was parameterized as a piecewise-constant profile with 45 four-minute intervals. The optimal open-loop profile is rather close to the optimal closed-loop trajectory (Figure 14.4). As a noticeable deviation is observed only toward the process end, it can be concluded that the process was not heavily perturbed by the process disturbances, and therefore, optimization with the nominal moment model could yield a reasonable solution. In Figure 14.5, the

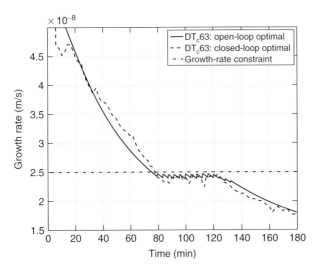

Figure 14.5 Simulated vs. actual crystal growth rate of DT_c63.

growth rates of the simulated open-loop optimal profile and the actual growth rate during batch DT_c63 are compared. As the profiles are in a good agreement, the open-loop implementation of the optimal trajectory could perform as good as the closed-loop strategy provided that the process would not experience any significant disturbances.

The agreement between the open-loop and the closed-loop trajectories becomes worse if there is an uncertainty in the initial conditions. The initial conditions of the reference batch DT_c31 were used to obtain an open-loop optimal profile of the heat input. Due to a difference in the initial conditions, the outcome of dynamic optimization diverges significantly from both the open-loop and the closed-loop optimal profiles of batch DT_c63 (Figure 14.4). Since the process disturbances of the considered process are weak, the difference in the initial conditions can be eliminated if they are estimated from the data obtained at the beginning of the batch and used for the process optimization. Given that optimization of the process performance does not consume much time, an optimal heat-input trajectory can then be computed off-line and implemented in the same batch in an open-loop fashion to obtain results that are similar to those of the closed-loop optimal control. The described strategy is similar to the so-called batch-to-batch variation control that is often used when online measurements of the CSD are not available (Ge et al., 2000).

14.3
MPC for Batch Crystallization

The model predictive control introduced in Chapter 16 can also be applied to batch crystallization. MPC can be used for online computation of an optimal input trajectory that minimizes a certain performance index subject to constraints. However, because of limitations imposed by the model used in the MPC controller or a more complicated objective function that cannot be adequately represented by the MPC cost function, alternative methods can be used for designing an optimal path. An off-line or an online optimization algorithm having a proper objective function can be used to generate optimal trajectories.

In the cases when model uncertainties and process disturbances are negligibly small, MPC can be used for trajectory tracking without an online dynamic optimizer as illustrated next. The optimal trajectory can be obtained through off-line optimization studies. It is worth mentioning that though online dynamic optimization is not used to update the optimal trajectory online, the MPC controller does perform optimization to achieve an optimal tracking of the given optimal profile. Because of the linearity of the relationship between the heat input and the crystal fraction in the crystallizer, a single linear model can be used in the classical MPC formulation as opposed to the applications with a strong nonlinear behavior, where multiple linear models are required (Shen, Chiu, and Wang, 1999). The relation between the heat input and the crystal growth rate is nonlinear as it depends

14.3 MPC for Batch Crystallization

on the mass and the surface area of the growing crystals. This relationship was approximated by another linear relationship.

The moment model presented in Section 13.8 was used to derive a finite impulse response model around the nominal operating trajectories. A process response to step tests was simulated with the moment model to obtain the data for model identification. The model identification was carried out using INCA modeler – a tool for identifying black-box models (IPCOS BV, the Netherlands). The identified models were found to be sufficient in describing the input–output relationship between the heat input on the one hand and the crystal fraction and the growth rate on the other hand. The models were used in the MPC controller (INCA®, IPCOS BV, the Netherlands). The delta-mode MPC (see Figure 16.7) was used without an online dynamic optimizer yielding a reference trajectory. The reference trajectory was given as a set-point whose value corresponded to the maximum achievable value of the crystal fraction at the final batch time as obtained by the off-line optimization studies (Section 13.8). The set-point was interpreted by the MPC controller as an increment to the values of the nominal trajectory. The nominal trajectory of the crystal fraction was the trajectory of a batch to which a constant input of 9 kW was applied (batch DT_c31).

The desired performance of the MPC controller was achieved by tuning the weights of the performance index of the delta-mode MPC optimization algorithm. By doing so, an implicit optimization similar to that by expressions in Equations (13.21) and (13.22) was implemented. The signals of the growth rate and the crystal fraction were used in the objective function formulation. The crystal fraction was calculated from the measured product density signal using Equation (13.20). The crystal growth rate was computed using Equation (13.19) in which the signal of the solute concentration was estimated as described in Section 14.2.

The MPC controller was implemented on the crystallization process (Figure 13.5) according to the philosophy described in Section 16.9 (see Figure 16.5). The above-mentioned observer (Section 14.2) was used as a soft sensor to monitor changes in the supersaturation and the crystal growth rate.

The results of an online test DT_c62 with the delta-mode MPC are shown in Figures 14.6–14.9. The process was operated with the heat input at its lower limit of 9 kW until the growth rate dropped below the maximum growth limit at about 80th min (see Figure 14.7). At that time, the MPC controller was switched on. The controller was operating with the initial values for tuning parameters until about 100th min, when it became obvious that the large weight was imposed on the deviation of the crystal fraction from the given set-point. Changing the tuning parameters of the controller resulted in a smooth but sluggish process response. This sluggishness manifested itself in large deviations from the desired crystal growth rate G_{max} (see Figure 14.7). The online tuning of the controller performance was clearly necessary.

The online tuning of the MPC controller was performed by changing the weights on the deviations of the controller variables from their respective set-points. The controller variables were the crystal volume fraction and the crystal growth rate. During the period on the unconstrained operation (see Figure 14.2), it was

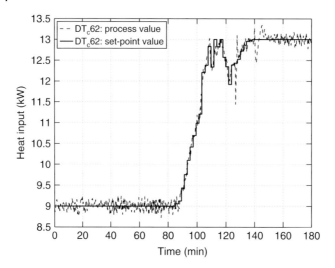

Figure 14.6 Optimal vs. implemented profile.

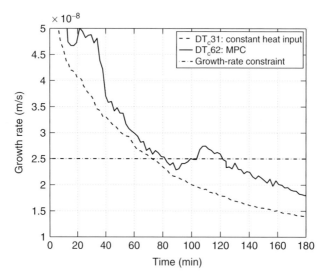

Figure 14.7 Growth rate during batch DT_c62.

necessary to maximize the crystal yield while keeping the growth rate as close as possible to its maximum value without exceeding it. As it was explained in the previous paragraph, the large penalty on the crystal-fraction error led to an aggressive controller response (see Figure 14.6). The resulting fast increase in the heat input caused the growth rate to overshoot its maximum value at about 100th min. Increasing the weight of the growth-rate error made the growth rate more important than the crystal fraction. As a result, the heat input was gradually lowered to bring the growth rate below the maximum value. Once the growth rate

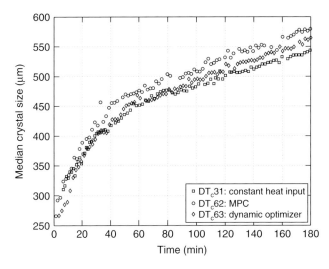

Figure 14.8 Comparison of the median sizes.

dropped below G_{max} at about 121st min, the heat input was being increased again to optimize the process yield. When the upper bound of 13 kW was reached, the controller lost its degree of freedom and the growth rate could no longer be kept at the set-point (Figure 14.7).

Both the dynamic optimizer and the MPC controller require tuning of their performance. The online tuning procedure described above could have been performed off-line with the process model acting as a process simulator. The off-line tuning of the closed-loop performance of the control system is recommended to ensure the process safety and avoid losses in productivity and/or the product quality. Online fine tuning is usually necessary to make final minor adjustments to the system behavior.

The maximum growth rate constraint was defined as a set-point/soft constraint. Its value was carefully chosen, so it can be exceeded for a short period without a considerable influence of nucleation on the product quality. Therefore, neither the median crystal size (Figure 14.8) nor the CSD width were significantly affected by the overshoot happened between the 100th and 121st min (Figure 14.7).

In Figures 14.8 and 14.9, the performance of the two advanced model-based techniques is compared in terms of the product quality and the crystal yield. It is obvious that both control methods perform better than the basic recipe control that was applied on batch DT_c31. A higher productivity of the process was obtained without compromising the product quality. In fact, the median crystal size was even better than that in the batch with a constant heat input as larger crystals are usually desired. The final width of the CSD was comparable in all considered processes.

As far as the comparison of the advanced techniques is concerned, there is hardly any difference in the process yield achieved in both operations. A slightly larger

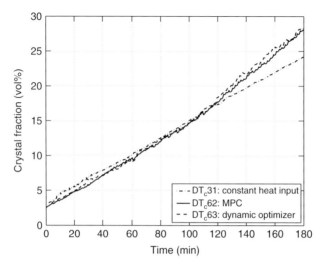

Figure 14.9 Comparison of the process yields.

median crystals size obtained in batch DT_c62 is a result of a higher supersaturation at the beginning of the batch (cf. Figures 14.3 and 14.7, Table 14.1).

To summarize, linearity of the relationship between the heat input and the crystal fraction did not allow for full exploration of the differences between the model-based techniques. The two control methods performed equally well. A distinct advantage of the dynamic optimizer is the direct use of the nonlinear process model. However, depending on the employed optimization algorithm, the use of such a model for numerical optimization can be computationally intensive. On the other hand, the use of linear models in the controller formulation can be computationally less demanding.

14.4
Conclusions and Perspectives

As it is shown in the above examples, the closed-loop implementation of the model-based control strategies is superior to the open-loop optimal control. Online computation of the input trajectories with feedback of the estimated process states to the optimizer/MPC controller enables early detection and rejection of the process disturbances and correction for the process–model mismatch, measurement errors, and the uncertain initial conditions. An optimal operation is achieved through a reduction in the variance of the controlled variables and a closer constraint tracking.

As industrial processes are rarely free of uncertainties, disturbances, and measurement noise, closed-loop control strategies are preferred over open-loop ones. Since the model-based control techniques rely on the process model, validity and accuracy of the models are crucial for the performance of the control system.

Recent developments in modeling show that effects of complex processes such as crystal attrition and breakage, agglomeration, and mixing can be described to assist in process design, optimization, and control studies. The advanced models become better in predicting the process behavior and therefore, they gain a potential for the use in the process scale-up.

Advances in development of online sensors for measuring the solute concentration, the CSD, and the crystal shape (Bakeev, 2005; Yu *et al.*, 2004) open up new perspectives for the process analysis, optimization, and design of effective process control strategies. Availability of the advanced sensors is essential for increasing the knowledge of the crystallization processes that is vital for establishing relationships between the process conditions and the crystal purity, morphology, and size, validating the modeled relationships, and optimizing the process control strategies. Reliable online measurements of the process state variables are also required for the implementation of the feedback control.

Finally, improvements of the simulation and optimization techniques facilitate obtaining an accurate, robust, and fast optimal solution (Kadam *et al.*, 2007). Fast optimization algorithms and advanced solution strategies are indispensable for online dynamic optimization of the process operation.

References

Agachi, P., Nagy, Z., Cristea, M., and Imre-Lucaci, A. (2006) *Model Based Control. Case Studies in Process Engineering*, Wiley-VCH Verlag GmbH & Co. KGaA, Weinheim.

Bakeev, K. (ed.) (2005) *Process Analytical Technology. Spectroscopic Tools and Implementation Strategies for the Chemical and Pharmaceutical Industries*, Blackwell Publishing, Ltd, Oxford.

Bonvin, D., Srinivasan, B., and Ruppen, D. (2002) Dynamic optimization in the batch chemical industry, in *Proceedings of the Sixth International Conference on Chemical Process Control, January 7–12, 2001, Tucson, Arizona*, AIChE Symposium Series No. 326, vol. 98 (eds J. Rawlings, B. Ogunnaike, and J. Eaton), AIChE: CACHE Publications, Austin, TX, pp. 255–273.

Ge, M., Wang, Q.-G., Chiu, M.-S., Lee, T.-H., Hang, C.-C., and Teo, K.-H. (2000) An effective technique for batch process optimization with application to crystallization. *Trans. IChemE*, **78A**, 99–106.

Kadam, J., Schlegel, M., Srinivasan, B., Bonvin, D., and Marquardt, W. (2007) Dynamic optimization in the presence of uncertainty: from off-line nominal solution to measurement-based implementation. *J. Process Control*, **17**, 389–398.

Kalbasenka, A.N. (2009) Model-based control of industrial batch crystallizers: experiments on enhanced controllability by seeding actuation. PhD thesis. Delft University of Technology, The Netherlands. ISBN: 978-90-6464-361-3. http://www.dcsc.tudelft.nl/Research/PhDtheses/ (accessed 2011)

Kalbasenka, A., Spierings, L., Huesman, A., and Kramer, H. (2007) Application of seeding as a process actuator in a model predictive control framework for fed-batch crystallization of ammonium sulphate. *Part. Syst. Charact.*, **24** (1), 40–48.

Shen, J.-X., Chiu, M.-S., and Wang, Q.-G. (1999) A comparative study of model-based control techniques for batch crystallization process. *J. Chem. Eng. Jpn.* **32** (4), 456–464.

Shi, D., El-Farra, N., Li, M., Mhaskar, P., and Christofides, P. (2006) Predictive control of particle size distribution in particulate processes. *Chem. Eng. Sci.*, **61**, 268–281.

Vega, A., Diez, F., and Alvarez, J. (1995) Programmed cooling control of a batch crystallizer. *Comput. Chem. Eng.*, **19**, S471–S476.

Zhang, G. and Rohani, S. (2003) On-line optimal control of a seeded batch cooling crystallizer. *Chem. Eng. Sci.*, **58**, 1887–1896.

Zhang, G. and Rohani, S. (2004) Dynamic optimal control of batch crystallization processes. *Chem. Eng. Commun.*, **191**, 356–372.

Yu, L., Lionberger, R., Raw, A., D'Costa, R., Wu, H., and Hussain, A. (2004) Applications of process analytical technology to crystallization processes. *Adv. Drug Deliv. Rev.*, **56**, 349–369.

15
Fines Removal
Angelo Chianese

15.1
Introduction

A very important target of crystallization processes is to produce a particulate solid product with a well-defined crystal-size distribution (CSD). Among CSD requirements, one of the most strict issue concerns the mass fraction of fines crystals, which are the crystals smaller than a fixed size. This specification is due to both the needs of merging the commercial properties of the final crystalline product and to decrease, as much as possible, the difficulties of crystallization downstream operations, as filtration, centrifugation, drying, and packaging.

To reduce the amount of fines sometimes is not possible at all, in other cases it is allowed by strongly changing the operating conditions and/or the type of the apparatus adopted for the crystallization process, but with dramatic cost increasing. In these cases, a useful choice to get the CSD target is to implement a control scheme of the crystallization process based on the fines removal.

15.2
Fines Removal by Heat Dissolution

The easiest way to remove fines is to warm a fines suspension stream up to a temperature at which all the fines are dissolved. The prerequisite of this operation is to separate the fines from the larger, that is, coarse, crystals in order to undertake only the dissolution of the classes of crystals to be removed from the final product. Therefore, the crystallizer has to be provided by a specific section where the coarse crystals are segregated from the fines, and then a mother liquor stream, where only fines are suspended, is withdrawn from the crystallization vessel, and sent to a system for the dissolution of fines. Finally, the mother liquor leaving the dissolution section is fed back to the crystallizer. The amount of fines destroyed in such a way is a function of both the magma density and the flow rate of the mother liquor stream with fines suspension.

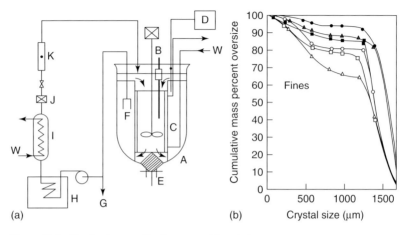

Figure 15.1 The batch crystallization setup (a) and the cumulative crystal size distribution (b) during batch crystallization of potassium sulfate crystals (Jones et al., 1984). Symbols: sphere (500 rpm), square (550 rpm), triangle (600 rpm), full with fines dissolution, empty without fines dissolution.

One of the first attempt to show the effectiveness of the fines destruction to increase the crystal size was reported in the works of Jones et al. (1984) and Jones and Chianese (1988) concerning the batch crystallization of potassium sulfate from aqueous solutions. The adopted set-up and the obtained results are reported in Figure 15.1. The circuit of the fines stream dissolution is the typical one to be implemented at lab scale: the crystals are classified in a fines trap (FT), transported by a pump to the inner part of an external heated tube and finally trough a cooling device. The latter is adopted in order to reduce the heating impact of the recycle stream on the crystallizer.

Very different cumulative CSDs were obtained by adopting an increasing flow rate of the fines stream sent to the dissolution device. In such a way, it was possible to reduce the mass percentage of fines in the best case from 24% to 9%.

When the size of the crystals sent to dissolution is of several tenths of microns the fines dissolution is not complete and a large thermal gradient between the heating fluid and the fines stream is required, as shown by the experimental work of Stoller et al. (2008). In this work, a fines stream of potassium sulfate crystals, smaller than 300 μm, was dissolved in an externally heated tube, with an inner diameter of 2 mm. The crystal suspension had a density of 50 g/l and temperature values of the heating oil were 130 °C and 150 °C, respectively. For a residence time of the crystal suspension of 3 s, a mass dissolution of crystals of 60% and 80% of the initial mass were observed, for oil temperature of 130 °C and 150 °C, respectively. The authors, by applying a CFD-based model, showed that 90% of the mass dissolution occurs inside the volume close to the tube wall. Therefore, in order to have a quick crystal destruction it is very important to attain a high temperature of the suspension in contact with the tube wall.

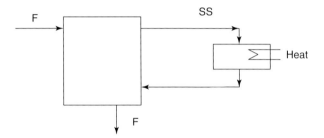

Figure 15.2 MSMPR crystallizer with fines removal and fines destruction.

15.3
Modeling of an MSMPR Continuous Crystallizer with Fines Removal

The effect of the fines removal operation on CSD may be easily modeled in the case of a mixed suspension mixed product removal (MSMPR) crystallizer operating in the continuous mode.

In this case, the scheme of the apparatus may be represented as shown in Figure 15.2, where F and SS are the volumetric flow rate of the feed stream and the fines crystal slurry stream, respectively, withdrawn from the crystallizer and sent to the dissolution section and L_F is the cut size of the removed fines.

All the fines crystals of size less than L_F leaves the crystallizer as solid suspension in the streams F and SS, and if all the fines present in the removed suspension SS are destroyed by heating, their residence time, τ_F, is equal to

$$\tau_F = \frac{V_C}{F + SS} \tag{15.1}$$

whereas the residence time of the coarse produced crystals, τ_P, is

$$\tau_P = \frac{V_C}{F} \tag{15.2}$$

For both the classes of crystals it is possible to apply the well-known population density equation written for an MSMPR crystallizer operating in the continuous mode, that is

$$n(L) = n_0 \cdot e^{-\frac{L}{\tau \cdot G}} \tag{15.3}$$

By applying Equation (15.3) to the population density of fines, $n_F(L)$, and coarse crystals, $n_P(L)$, the following expression are obtained, respectively:

$$n_F(L) = n_0 \cdot e^{-\frac{L}{\tau_F \cdot G}} \tag{15.4}$$

and

$$n(L) = n_{0,P} \cdot e^{-\frac{L}{\tau_P \cdot G}} \tag{15.5}$$

It has to be noticed that the population density at $L \to 0$ for the fines crystals size distribution, n_0, is a physical value derived from the nucleation rate, whereas $n_{0,P}$ is a hypothetical value-giving rise to CSD of coarse crystals. This latter value may

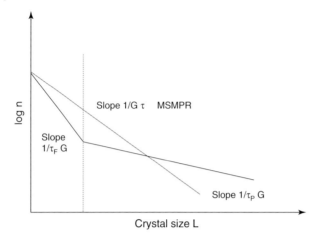

Figure 15.3 Population density vs. crystal size for an MSMPR crystallizer with fines removal.

be derived from the condition that the two Equations (15.4) and (15.5) lead to the same population density value in correspondence of $L = L_F$, that is

$$n_0 \cdot e^{-\frac{L_F}{\tau_F \cdot G}} = n_{0,P} e^{-\frac{L_F}{\tau_P G}} \qquad (15.6)$$

In Figure 15.3, CSD of fines and coarse crystals are reported in the semilogarithmic plot log $n(L)$ versus L.

By putting in Equation (15.5) the expression of $n_{0,P}$ derived from Equation (15.6) the equation representing the population balance in the whole crystal size range is obtained (Randolph and Larson, 1988):

$$n(L) = n_{0,F} \cdot e^{-\frac{L_F}{\tau_F \cdot G}} e^{-\frac{L-L_F}{\tau_P G}} \qquad (15.7)$$

For sake of comparison, in Figure 15.3 it is also plotted the population density for an MSMPR crystallizer without fines removal. It is evident that the technique of fines removal in principle leads to the higher value of the mass ratio coarse/fines crystals and of the crystal product average size, as a consequence.

15.4
Fines Destruction in the Industrial Practice

The typical crystallizer adopted to allow a fines dissolution scheme is the so-called draft tube baffle (DTB) crystallizer (see Figure 15.4). This crystallizer provides for two discharge streams: the outlet slurry that contains the product crystals, from the bottom, and the outlet fines stream, consisting of mother liquor with a small amount of suspended fine crystals, from the upper right side. This latter stream is recirculated to the bottom of the crystallizer through a heat exchanger that warms up the stream at a temperature at which all the crystals are dissolved. The DTB

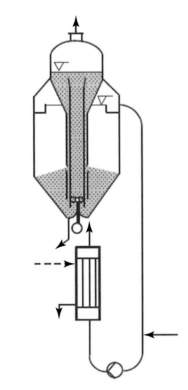

Figure 15.4 DTB crystallizer provided with a heat exchanger in the external circuit.

crystallizer provides the production of crystals with a narrow-size range distribution and a quite high average size up to 1.5 mm.

The fines dissolution process suffers from many constraints.

1) For an operating crystallizer, the fines stream flow rate is fixed and determined by the cut size of fines to be removed from the annular section of the crystallizer.
2) For cooling crystallizers, the temperature of the recycling stream exiting the heat exchanger has to be high enough to assure a complete dissolution of the fines, but not so high to strongly increase the heat content of the recycled warm stream.
3) For evaporative crystallizers, this latter aspect is not a problem, but, since the heat exchanger is mainly used to provide the evaporation heat, there is no chance to adjust the heat duty for fines dissolution control purpose and the danger of the dramatic phenomenon of cycling may occur.

The measurement of the fines content in the stream outgoing the upper part of the crystallizer should be advisable for monitoring and control purposes, but it is a difficult task because of the shortage of online sensors accurate and at a reasonable price.

For all the above-mentioned constraints there are few examples of CSD feedback control for an industrial crystallizer by using the amount of dissolved fines as a manipulated variable. In other words, the amount of fines in the outlet slurry from a DTB crystallizer is reduced just by its design configuration, without applying a CSD feedback control scheme.

Otherwise, in the literature, there are many examples of CSD automatic control schemes applied to pilot plant crystallizers and some of them are reported in the following section.

15.5
CSD Control by Fines Removal for Pilot Scale Crystallizers

One of the first work, giving an example of application of fines removal for a pilot plant crystallizer, is the one published by Randolph, Chen, and Tavana (1987). In this work an online CSD control in an 18 l KCl crystallizer, equipped with a fines dissolver, was experimentally studied using inferential control of nuclei density. This latter variable was estimated by using a light scattering instrument. Simple proportional control of nuclei density in response of changes of fines removal rate was shown to be an effective way to minimize fluctuations in the product CSD that are caused by nucleation disturbances. The fines stream was removed from an internal FT of the crystallizer where coarse crystals segregation takes place. An important aspect of the experimental apparatus was that the flow rate of the fines dissolver stream was adjusted by recycling a portion of it directly back to the crystallizer, in order not to vary the cut size of the fines being removed from the crystallizer itself.

Some years later, Rohani and Paine (1991) proposed a new measurement system of the fines content; it consists of a dissolution cell, provided with an external thermostated jacket, where a fines suspension sample was progressively heated until the last crystal disappeared. The disappearance of the solid phase was detected by means of a turbidimeter. This sample and effective instrument was used to control the fines content in a KCl continuous cooling crystallizer.

Russeau and Barthe (2005) applied the fines destruction technique to control paracetamol CSD in a batch cooling crystallizer. An FBRM was used to monitor the evolution of CLDs. The results demonstrate how selective fines destruction influences the size distribution of the crystalline products.

To control the potassium sulfate CSD from a pilot plant cooling crystallizer, Bravi and Chianese used a sophisticated turbidimetric cell to measure the content of fines crystals and adopted a neuro-fuzzy-control algorithm (see Figure 15.5) (Bravi and Chianese, 2003).

The fines slurry stream, at a constant flow rate, was withdrawn from the crystallizer through a FT. The adopted flow rate led to the segregation of large crystals and just fines smaller than 250 μm in size were removed. The fines stream was pumped to the fines density measuring instrument (FDMI) (Chianese, Bravi, and Kuester, 2002) flow cell. Then the VP-3 volumetric pump was operated to transfer a fraction of the fines stream to sedimentation-dissolution unit T-3, fitted with a thermal resistance, where most of the crystals settle and separate from the liquid solution, while the crystals smaller than 40 μm were entrained and dissolved

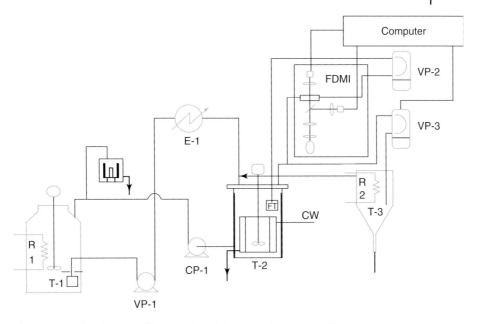

Figure 15.5 Pilot plant crystallizer for the production of potassium sulfate.

in the upper part of the tank by the heating. The crystallizer outlet slurry stream was recycled back to the crystallizer tank.

The main operating parameters (temperatures and flow rates) of the pilot plant were acquired by a personal computer through a data acquisition and control interface card. The control tool included a fuzzy algorithm, written in the MATLAB "simulink" environment, a low-pass filter to eliminate the noises and a plant supervision unit.

The online measurement of the particle concentration was performed by means of online turbidimeter, essentially consisting of a laser beam passing across an optical cell where the crystal suspension flows through. The adopted neuro-fuzzy controller was the synthesis of a fuzzy controller and a neural adaptive system for the optimal shaping of its membership functions.

Figure 15.6 reported the process dynamic of the examined system in terms of concentration of fines from the run start until the operation was put under control of the fines content. The run start was considered to correspond to the first nucleation. Afterward, the mass of fines increased until they grew to a larger size-giving rise to a peak in the fines concentration. After 4 h a steady-state condition was attained and it was possible to start the control action to maintain the fines concentration at a constant value of 8 kg/m^3. At a time of 4000 s, the set-point of the fines controller was quickly fixed at 3.5 kg/m^3. The crystallizer operation was forced respect to the new target, which was reached after a rise time of 700 s. Finally, after a limited oscillation a stable new steady state was achieved. This way the amount of fines, equal to 29.2%, at the initial steady state, was decreased to around 20% at new operating conditions.

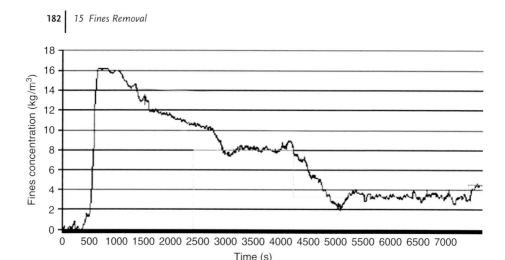

Figure 15.6 Time history of the fines concentration during a typical servomechanism-problem run under neuro-fuzzy control.

15.6
The Cycling Phenomenon as Undesired Effect of Fines Destruction in Industrial Crystallizers

In the absence of a control scheme, the removal of fines from a DTB crystallizer by heat dissolution may promote in industrial crystallizers the phenomenon of cycling, which consists of a dramatic oscillation of size distribution of the crystal product from an industrial crystallizer with a period of time of some tens of hours. Figure 15.7 shows typical examples of cycling taking place in an evaporative ammonium sulfate crystallizer, provided with a fines destruction loop. Cycling takes place when the rate of fines dissolution by heating is higher than the secondary nucleation rate within the crystallizer. In such a case, the amount of fines progressively decreases and the crystal surface prone to growth, as well. As a consequence, the supersaturation, generated by solvent vaporization, is not transformed to deposited mass and progressively increases, until primary heterogeneous nucleation occurs with the production of a big number of fines. When the presence of fines in the crystallizer decreases, due to the fines dissolution device, the growth rate of the coarse crystals largely increases and large crystals, few millimeters in size, are obtained.

The sequence of crystal population images throughout cycling is reported in Figure 15.8.

At the industrial level, it is preferred to remove cycling by changing strategy. Van Esch et al. (2008) proposed to continuously seed the crystallizer in order to make always available a crystal surface where supersaturation may be discharged on. This strategy allowed to eliminate cycling phenomenon and to obtain a quite stable crystallizer operation.

15.6 The Cycling Phenomenon as Undesired Effect of Fines Destruction in Industrial Crystallizers

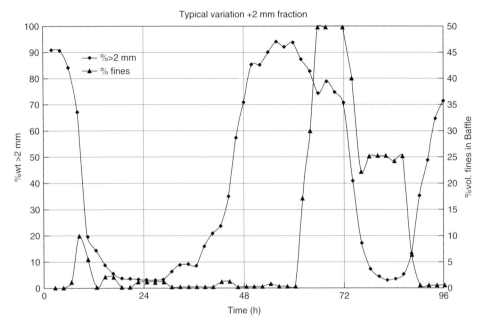

Figure 15.7 Evolution of coarse and fines crystals during cycling for an industrial crystallizer. (Courtesy of GEA Messo Italy.)

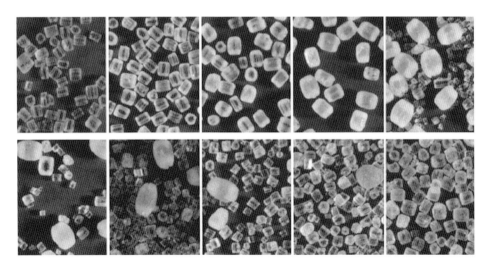

Figure 15.8 Ammonium sulfate crystal populations throughout cycling. (Courtesy of GEA Process Engineering Italy.)

References

Bravi, M. and Chianese, A. (2003) Neuro-fuzzy control of a continuous cooled crystallizer. *Chem. Eng. Technol.*, **26**, 262–266.

Chianese, A., Bravi, M., and Kuester, C. (2002) A crystal size related feature measurement instrument for the process industry. *Chem. Eng. Trans.*, **1**, 1491–1496. Proceedings of 15th Symposium on Industrial Crystallization, Sorrento, September 15–18, 2002.

Jones, A.G. and Chianese, A. (1988) Fines destruction during batch crystallization. *Chem. Eng. Commun.*, **62**, 5–16.

Jones, A.G., Chianese A., and Mullin, J.W. (1984) *Industrial Crystallization 84*, Elsevier, The Netherlands, pp. 191–194.

Randolph, A.D., Chen, L., and Tavana, A. (1987) Feedback control of CSD in a KCl crystallizer with fines dissolver. *AIChE J.* **33** (4) 583–591.

Randolph, A.D. and Larson, M.A. (1988) *Theory of Particulate Processes*, 2nd edn, Academic Press, San Diego, CA.

Rohani, S. and Paine, K. (1991) Feedback control of crystal size distribution in a continuous cooling crystallizer. *Can. J. Chem. Eng.* **69**, 165–171.

Rousseau, R.W., Barthe, S. (2005) Using FBRM measurements, fines destruction and varying cooling rates to control paracetamol Csd in a batch cooling crystallizer, AIChE Annual Meeting, Conference Proceedings, p. 2279

Stoller, M., Orlandi, P., Leonardi, S., and Chianese, A. (2008) Modelling of the fine crystals dissolution in an externally heated tube. Proceedings of 18th Symposium on Industrial Crystallization, Vol. 2, Maastricht, September 14–17, 2008.

Van Esch, J., Fakatselis, T.E., Paroli, F., Scholz, R., and Hofmann, G. (2008) Ammonium sulfate crystallization – state of the art and trends. Proceedings of 18th Symposium on Industrial Crystallization, Vol. 2, Maastricht, September 14–17, 2008, pp. 859–865.

16
Model Predictive Control
Alex N. Kalbasenka, Adrie E.M. Huesman, and Herman J.M. Kramer

16.1
Introduction

Model predictive control (MPC) is a collective name for a range of supervisory control methods that have the following common features. Supervisory control is a part of the overall control system of a process unit. The overall control system can be decomposed into two levels: a base control layer and an advanced or supervisory control level. The base control layer is composed of single-loop controllers operating in the range of seconds. The supervisory controllers reside in the advanced control layer and operate in the range of minutes. The controlled unit is usually a dynamic multivariable process with a plurality of requirements. The requirements range from operating and product quality-related issues to economic objectives. The hierarchy of the requirements is characterized by the difference in their importance. The MPC controller attempts to satisfy the most important requirements first. Then, if some degrees of freedom are left, less important requirements are considered by the predictive controller.

The predictive control methods use a process model to predict the behavior of process outputs based on the information on the past inputs and outputs as well as on future input trajectories (Figure 16.1). The future input trajectories are obtained as a solution to a minimization problem using an optimizer. The optimizer minimizes a cost function (also called the *performance index*) subject to constraints on the input, the output, and the input increment trajectories. The cost function should not be confused with the economic costs incurred by the process.

16.1.1
Receding Horizon Principle

The MPC controller operates using the so-called receding horizon or moving horizon principle. The moving horizon principle can be explained with the help of Figures 16.1 and 16.2 and the performance index in the next simplified form

$$J = \sum (\hat{\mathbf{y}} - \mathbf{y}_{\text{ref}})^T \mathbf{Q} (\hat{\mathbf{y}} - \mathbf{y}_{\text{ref}}) + \sum \Delta \mathbf{u}^T \mathbf{R} \Delta \mathbf{u} \qquad (16.1)$$

Industrial Crystallization Process Monitoring and Control, First Edition. Edited by Angelo Chianese and Herman J. M. Kramer.
© 2012 Wiley-VCH Verlag GmbH & Co. KGaA. Published 2012 by Wiley-VCH Verlag GmbH & Co. KGaA.

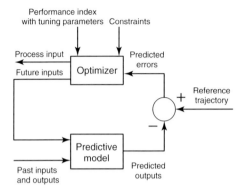

Figure 16.1 Structure of an MPC algorithm.

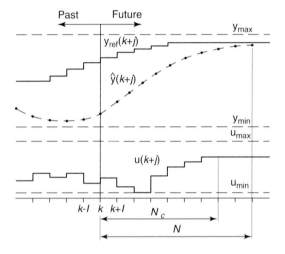

Figure 16.2 Receding horizon principle.

in which indices were omitted for simplicity. Using the predictive model, the predicted outputs $\hat{y}(k+j)$ for $j = 1, \ldots, N$ are calculated on the prediction horizon N. The information on the past inputs and outputs as well as on the optimal future control actions $u(k+j-1)$ for $j = 1, \ldots, N_c$ on the control horizon N_c is used for the output prediction.

The future optimal input trajectory is calculated by minimizing the performance index J (in the form of, for instance, Equation (16.1)). The goal is to keep the deviations $(\hat{y} - y_{\text{ref}})$ of the predicted process outputs from the reference trajectory (or a set-point) and the increments Δu of the input signal u as small as possible respecting all the imposed constraints. To indicate a higher importance of a certain input increment or output signal, a larger weight is assigned to the corresponding element of the diagonal weighting matrix $\mathbf{R}$ or $\mathbf{Q}$. Therefore, minimization of the performance index amounts to trading off between trajectory tracking and minimization of the control effort needed for bringing the process outputs to the reference signals. If the requirements on the control effort are stringent, accurate tracking of the reference signal might be not reached. And vice versa, if the required tracking properties are such that most of the degrees of freedom are spent on it, minimization of the energy of the inputs might be impaired. To summarize, depending on the weights and degrees of freedom available, the control objectives can be satisfied partly or completely.

The optimal solution of the minimization problem is a sequence of the future inputs at which the performance index takes its minimal value. The first element $u(k)$ of the optimal control sequence is implemented as the process input. The prediction horizon shifts one sample ahead and the calculation procedure is repeated.

16.1.2
Advantages and Disadvantages of MPC

The MPC technology was developed in 1970s, and since then found many applications in industries. The success of the MPC is due to a number of advantages over other control techniques.

One of the main advantages of MPC is that it uses a model to predict the process outputs and to compute the future control sequence that keeps process variables between the bounds. The latter can be introduced by means of constraints on the process inputs and outputs. MPC handles constraints explicitly while optimizing the process performance index. In other words, the constraints are directly taken into account during the optimization. Therefore, the problem of integrator wind-up occurring during the performance of other controllers can be avoided. Classical control methods, such as proportional-integral-derivative (PID) control, do not consider consequences of the current process inputs and take the action based on the past errors only.

In the case of control of multiple-input multiple-output (MIMO) systems, PID controllers do not take interactions among different control loops into consideration. Therefore, the design of a control system for such multivariable systems using traditional PID control can be a challenge. MPC methods make use of models to account for a complex interplay amongst the process variables. A trade-off in the performance of the different loops can be accounted for in a systematic way by defining a proper cost function.

Since flexibility in process operation is a prerequisite for meeting the dynamic market demands, the process is required to operate in a wide range of process conditions. To ensure a profitable operation, the process performance requirements can rapidly change in response to the market conditions. The performance criteria for the process unit can then take one or a combination of the economic requirements (maximization of the process yield, minimization of the operating costs, and/or the processing time), product quality-related specifications (product grade, purity, and minimization of the transition time between the grades or an off-spec product produced during the transition), and environmental issues. The

MPC technology allows for flexibility with respect to the changing performance requirements and operating conditions without having to modify the controller structure.

Another benefit of MPC is that the objective function can be formulated in such a way as to minimize operating costs. The process operating condition can be then optimized with respect to an economic objective function. In fact, the objective function can take into account any economic measure. It has to be noted here that a direct implementation of such an optimization objective is limited by the form that the performance index (given by Equation (16.1)) has. If the desired objective cannot be formulated in terms of the model inputs and outputs alone, an off-line or an online optimization with a dedicated process model and objective function can be carried out. The resulting optimal output and/or input profiles can serve as reference trajectories for a delta-mode MPC controller (see Section 16.11 for details).

A prioritized formulation of a constrained optimization allows for flexibility in defining a control strategy. Ranking of the requirements formulated as set-points and bounds or zones for different variables is needed to reflect the degree of the variable importance. For instance, safety requirements have a higher priority than quality requirements. Therefore, the optimizer first attempts finding an optimal solution for requirements with a higher priority. If there are still some degrees of freedom left, the satisfaction of lower priority specifications is attempted next.

As an optimal process operation is often an operation that is close to a constraint, an explicit use of constraints by MPC can be exploited by pushing the average value of the controlled variables closer to the optimal constraint limit (Van Brempt et al., 2001). If the MPC controller is designed to have a wide bandwidth, the controller is capable of correcting for process disturbances and reducing the variance of the controlled variables. These properties of the controller allow for the process operation close to the optimality without constraint violations.

The requirement of the predictive model is at the same time a disadvantage of MPC. A model with good predictive capability is not always available and has to be derived. As modeling is one of the most important steps, it requires a substantial effort. The model accuracy determines the performance of MPC. In order to obtain an accurate prediction over a long prediction horizon, the dynamic model should be capable of describing the dominant process dynamics. This requirement implies availability of a rather complex process model. On the other hand, the model should be as simple as possible to make real-time optimization feasible. It is evident that accuracy and simplicity are often contradictory requirements. In order to improve the model prediction, state observers can be used to correct for process disturbances, uncertain initial conditions, measurement noise, and modeling errors. State estimators or observers are software sensors capable of reconstructing unmeasured variables from a limited set of measurements with the use of a process model (Dochain, 2003). The observers do improve the model prediction, however at the expense of a more complex design of the predictive control system.

16.2
Approach for Designing and Implementing an MPC Control System

Since a number of commercial MPC technologies are available (aspenONE™ Advanced Process Control Solution, AspenTech, USA; INCA®, IPCOS BV, The Netherlands; and some others listed in Agachi *et al.* (2006)) details concerning the actual implementation of the control algorithms are left out (the interested reader is referred to Camacho and Bordons (1999), Maciejowski (2002), Rossiter (2003) and Muske and Rawlings (1993)). Instead, general guidelines on how to design and implement an MPC control configuration using available commercial technology and research tools (such as the MPC Toolbox and System Identification Toolbox of MATLAB®, MathWorks, USA) are given.

A general procedure underlining major milestones in designing and implementing a MPC strategy is shown in Figure 16.3. The entire design procedure is divided into three main activities, namely the process modeling, the model-based control design, and the implementation, which are discussed next.

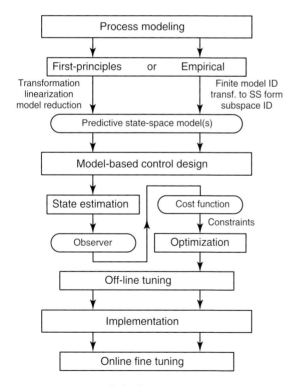

Figure 16.3 Approach for designing an MPC control system.

16.3
Process Modeling

As is clear from the name of the discussed control strategy, predictive models are essential for MPC. Models are used for prediction of the future behavior of the process outputs on the basis of the input signals applied in the past, information about past, and expected process disturbances. The predicted future outputs are used for calculation of the process inputs that are optimal with respect to the given cost function, specifications on the controlled variables (set-point values and bounds), and constraints on the process inputs.

As the model quality determines the quality of the output prediction and therefore, the predictive control, modeling is a crucial step in developing the control system. While it is desirable to have a model that is as simple as possible, the relevant process dynamics has to be described accurately. There are various modeling methods to obtain a suitable process model. These methods can be categorized into two categories: first-principles modeling and empirical modeling. In principle, the predictive model can be of any nature and form. However, for variety of reasons linear state-space (SS) models are preferred (compactness, suitability for MIMO systems, availability of classical linear control theory for design and analysis of a controller and a state estimator, and possibility of converting other linear models to a SS form).

Since the predictive control problem is implemented in a sampled-data control system and solved using a digital computer, the underlying process model has to be discrete. A linear discrete SS model has the following representation:

$$\mathbf{x}(k+1) = \mathbf{A}\mathbf{x}(k) + \mathbf{B}\mathbf{u}(k)$$
$$\mathbf{y}(k) = \mathbf{C}\mathbf{x}(k)$$
(16.2)

in which $\mathbf{x}(k)$ is the state vector (having dimension of $1 \times n$), $\mathbf{y}(k)$ is the vector of process outputs or controlled variables ($1 \times l$), and $\mathbf{u}(k)$ is the vector of inputs or manipulated variables ($1 \times m$) at sample time k. $\mathbf{A}$ ($n \times n$), $\mathbf{B}$ ($n \times m$), and $\mathbf{C}$ ($l \times n$) are the time-invariant matrices of the SS model.

When information on the process disturbance is available, a disturbance model should be included into the process model. The disturbance model allows description of the process behavior not reflected by the process model such as the effect of nonmeasurable disturbances, model errors, and process and measurement noise. A SS representation of the disturbance model can be added to the process model (Equation (16.2)) to yield an augmented process model.

First-principles modeling relies on fundamental conservation laws to model energy and mass (for the crystallized material in both the liquid and solid phases) balances. The crystal size distribution (CSD) can be modeled using the population balance equation (PBE). The resulting models are usually rather complex and nonlinear. The system consists of a set of coupled integro-differential equations (mass and energy balances) and a hyperbolic partial differential equation (PDE) (population balance). The solution techniques for such a system include transformation of the model PDE by using different techniques (method of moments, method of characteristics, finite difference methods, and finite element methods) into a set of (nonlinear) differential algebraic equations (DAEs). The resulting system of equations is usually nonlinear. Linearization can be used to obtain a linear model. The linearized model can be of a rather high order to be effectively used in real-time optimization. Therefore, a model reduction step can be applied to obtain a reduced linear model (Eek, 1995).

The crystallization kinetics is often modeled using empirical relationships with parameters whose values are determined on the basis of experimental data. Therefore, strictly speaking, the resulting overall crystallization model is of hybrid nature.

Linear models derived in such a way can be poor in describing nonlinear process dynamics as each subsequent step in the model development introduces an error. The first-principles or hybrid models can be used for derivation of linear models suitable for application in MPC in a different way that is discussed next.

Empirical modeling uses the process data to derive an input–output relationship. A disadvantage of empirical modeling is that a large number of experiments are required to obtain a dataset capturing the process dynamics. Performing the tests implies perturbations of the process inputs that can lead to loss of productivity and product quality. In that case, a validated first-principles model can be used for model identification (Van Brempt *et al.*, 2001; Rohani, Haeri, and Wood, 1999a). The first-principles model is validated using the process data originating from a historical database of the process or from dedicated tests that are usually much shorter than the identification tests. The validated model is then used to simulate the data needed for identification.

Identified linear models have usually a limited description capability of the process dynamics at different time scales. Therefore, it can be advantageous to combine model identification using a first-principles model and the process data to derive a composite linear model taking into account slow and fast process dynamics (Van Brempt *et al.*, 2001).

The description capability of an identified model is limited to the input trajectories used in identification experiments to excite the process dynamics. As a result, the output prediction for input trajectories that are outside of the assumed operational envelop can lead to large prediction errors. As there can be an enormous number of practical input trajectories, the choice of an optimal input signal for an identification test is critical. The most commonly used are pseudo-random binary signals (PRBSs) with switching interval as an adjustable parameter. Other adjustable parameters include the duration of the test, the amplitude of the steps, and the frequency of switching between the two amplitude levels used in the PRBS signal. The length of the switching interval should be of order of magnitude of the process-effective time constant. Rohani *et al.* (1999b) report using a random-magnitude random-interval (RMRI) signal as a better alternative to a PRBS signal as the former covers a larger frequency range.

In principle, any model capable of accurately predicting the behavior of the process outputs can be used in an MPC formulation. The most commonly used models are finite impulse response (FIR) models, step response models, transfer function models, and SS models. All these models can be transformed into each other. Each type of the models has a dedicated formulation of MPC (Camacho and Bordons, 1999; Rossiter, 2003; Maciejowski, 2002). Nonlinear models can also be used in MPC; however, this is not yet a general practice (Rohani, Haeri, and Wood, 1999a).

As mentioned above, an MPC formulation based on SS models can be preferred. Such a formulation is flexible with respect to the choice of a modeling method as many methods are capable of yielding a state-space model after necessary transformations.

As for model identification, subspace identification methods offer a simple and general parameterization for MIMO systems and allow for direct identification of a state-space model both for open-loop and closed-loop systems.

As most of the commercial MPC technologies provide a user with a model identification package, the details on the underlying principles of identification are omitted. The interested reader is referred to Ljung (1999) for the theory and methods of system identification. A review on subspace identification methods can be found in Qin (2006).

16.4
The Performance Index

The performance index or the cost function is a numerical measure that is used to compute an optimal future control sequence. The cost function is a mathematical description of the required process performance that can be expressed as a single number so that different input trajectories can be compared on the basis of their cost. The choice of the performance index determines not only the process performance, but also the complexity of the optimization problem. Therefore, quadratic cost functions are often used as they result in a straightforward optimization:

$$J = \sum_{j=N_m}^{N} \left(\hat{\mathbf{y}}\left(k+j\right) - \mathbf{y}_{\text{ref}}\left(k+j\right)\right)^T \mathbf{Q} \left(\hat{\mathbf{y}}\left(k+j\right) - \mathbf{y}_{\text{ref}}\left(k+j\right)\right)$$
$$+ \sum_{j=1}^{N_c} \Delta \mathbf{u}^T\left(k+j-1\right) \mathbf{R} \Delta \mathbf{u}\left(k+j-1\right) \quad (16.3)$$

where $\mathbf{Q}$ and $\mathbf{R}$ are positive definite diagonal weighting matrices, N is the prediction horizon, N_c is the control horizon, N_m is the minimum-cost horizon, $\mathbf{y}_{\text{ref}}(k+j)$ is the reference trajectory, $\hat{\mathbf{y}}(k+j)$ is the predicted output, $\Delta \mathbf{u}(k+j-1) = \mathbf{u}(k+j-1) - \mathbf{u}(k+j-2)$ is the vector of input increments.

An additional term including the values of control signal $\mathbf{u}(k+j)$ can be added to the performance index.

16.5
Constraints

Three types of constraints can be distinguished: input constraints, outputs constraints, and constraints on the rate of change of the process inputs. Constraints on the manipulated variables can be dictated by the physical limitations of the used hardware (limitations in pump capacity, heat transfer limitations, and control-valve saturation) or operational reasons. Constraints on the controlled variables are imposed when variations from a predefined set-point have to be limited or when the admissible outputs are defined by the upper and the lower bounds. The rate-of-change constraints are imposed on the process inputs when the calculated input sequence suggests too steep variations for the controller to handle or when the process behavior can be adversely affected by such changes (e.g., large levels of supersaturation can be achieved in a short period of time leading to an undesired nucleation). Mathematically, the three types of constraints can be written as

$$\begin{aligned} \mathbf{y}_{min} \leq \mathbf{y}(k+j) \leq \mathbf{y}_{max} & \quad j=1,\ldots,N \\ \mathbf{u}_{min} \leq \mathbf{u}(k+j) \leq \mathbf{u}_{max} & \quad j=1,\ldots,N_c \\ \Delta\mathbf{u}_{min} \leq \Delta\mathbf{u}(k+j) \leq \Delta\mathbf{u}_{max} & \quad j=1,\ldots,N_c \end{aligned} \quad (16.4)$$

16.6
The MPC Optimization

Finding an optimal input sequence means solving an optimization problem that minimizes the performance index J (Equation (16.3)) by varying N_c future control inputs. This minimization can be represented by

$$\min_{\{u(k),\ldots,u(k+N_C-1)\}} J \quad (16.5)$$

subject to model (Equation (16.2)) and constraints (Equation (16.4)).

The size of optimization problem (Equation (16.5)) depends on the number of process variables and the length of prediction and control horizons. The tuning parameters have an impact on the outcome of the optimization.

16.7
Tuning

The performance of an MPC controller with respect to signal tracking, disturbance rejection, and robustness to the mismatch between the model and the process can be influenced by adjusting the tuning parameters N_m, N_c, N, $\mathbf{Q}$, and $\mathbf{R}$. Though there is no systematic way of tuning the parameters on the basis of the desired process performance, some basic guidelines can be formulated for designing the cost function.

The minimum-cost horizon is usually chosen $N_m = 1$. However, if the process exhibits a dead time T_d or an inverse response, it is advised to assume $N_m = 1 + T_d$, where the dead time T_d is a multiple of the sample interval. Elimination of the first samples corresponding to the dead time or the inverse response from the objective function (Equation (16.3)) will result in a more smooth process response.

The length of the prediction horizon N is normally chosen equal to the length of the step response of the open-loop process. In other words, the prediction interval N should capture the crucial dynamics of the process outputs being the response to the process inputs. The length of the prediction horizon also depends on the model quality. The quality of prediction on a longer horizon is subject to the model accuracy. If the model is not accurate enough to anticipate the distant future, the process cannot be controlled optimally. The length of N has also implications on the computational time.

The control horizon N_c is usually taken such that $N_c < N$ with $\Delta \mathbf{u}(k+j) = 0$ for $j > N_c$. This implies that for all $j > N_c$ the input signal is kept constant and $\mathbf{u}(k+j) = u(k+N_c-1)$. Choosing a long control horizon can complicate finding an optimal solution. A shorter control horizon is beneficial not only from the viewpoint of stability, but also for reducing the computational effort for solving the MPC optimization.

The weighting matrices $\mathbf{Q}$ and $\mathbf{R}$ allow prioritizing performance of different loops. It is advisable to normalize all input and output signals in such a way that the range from 0 to 1 has an equal importance for all loops. Then, the initial values of the elements of matrices $\mathbf{Q}$ and $\mathbf{R}$ can be set to unity. Since a systematic approach on how to choose weights based on the desired performance characteristics is absent, fine-tuning has to be performed online.

16.8
State Estimation

Including a state estimator (observer) into the MPC control strategy allows for determining the current state of the process from available measurements and calculating the output predictions based on the estimated states (Froisy, 2006). Without correcting the process model by means of a feedback mechanism such as the one offered by an observer, the model predictions are likely to diverge from the measured plant outputs (Figure 16.4a). The future control moves computed using inaccurate model predictions are, therefore, not optimal. As the observer corrects for process disturbances, uncertain initial conditions, measurement noise, and modeling errors; the prediction of the process outputs becomes more accurate (Figure 16.4b).

As is shown by Eek (1995), a state estimator is essential for a good performance of MPC of a single-input single-output (SISO) system. A traditional PID control of the same system was superior to MPC without state estimation (see Section 16.10).

Apart from enhancing performance of the control system, an observer can be used for the purposes of process control, monitoring, and modeling. The observer

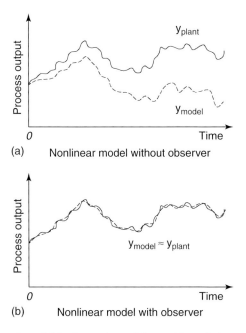

Figure 16.4 Comparison of the model prediction and the measured plant output.

is then designed to estimate a process parameter or a variable that otherwise cannot be measured due to unavailability, an insufficient accuracy, or a high cost of sensors. The principles of observation technology are briefly explained below.

For crystallization system described by a model in the following general form:

$$\dot{x}(t) = f(x(t), u(t)), \quad x(t_0) = x_0$$
$$y(t) = h(x(t))$$
(16.6)

a common representation of an observer can be given by

$$\dot{\bar{x}}(t) = f(\bar{x}(t), u(t)) + K(h(\bar{x}(t)) - y(t))$$
(16.7)

in which $\bar{x}(t)$ is the state vector estimated on the basis of available information on the process inputs $u(t)$ and the measured outputs $y(t)$. As is obvious from Equation (16.7), the state estimates are obtained by correcting the model predictions by the measurement errors $(h(\bar{x}(t)) - y(t))$. K is the observer gain. The gain is tuned in such a way as to obtain a desirable dynamics of the estimation errors $(\bar{x}(t) - x(t))$.

There are different types of observers whose description can be given by Equation (16.7). The differences among the observers are in the underlying assumptions and the methods that are used to compute the observer gain. The Extended Kalman filter (EKF) and the Extended Luenberger observer (ELO) are the two observers frequently used for nonlinear process systems such as crystallization. Details on the mentioned techniques and other available methods are given in Dochain (2003).

Examples of implementation of the mentioned observation techniques can be found in, for instance, Tadayyon and Rohani (2001) and Kalbasenka *et al.* (2006), and Kalbasenka (2009).

16.9
Implementation

The MPC controller should be implemented as a superstructure to an existing distributed control system (DCS) or programmable logic controller (PLC) system consisting of, for instance, traditional PID controller loops (Figure 16.5). Such an implementation gives flexibility with respect to the choice of the control strategy at different stages of a process. For instance, the DCS can be used to control the process at the start-up with the MPC controller being inactive. That is, the future inputs computed by the MPC algorithm are not used as set-points of PID controllers. The MPC controller can be activated at a later stage.

Another advantage of the MPC implementation in the supervisory control level is safety. If the MPC controller or the communication between the MPC controller and the DCS fails, the control of the process unit is taken over by the PID controllers of the base control layer.

The core of the architecture presented in Figure 16.5 is an OPC (object linking and embedding for process control) server. The OPC server enables exchange of information among different applications. So, the measured variables and the applied inputs registered by the DCS can be written to and subsequently read from the server by other control modules such as an observer and an MPC controller.

Commercial MPC products as well as prototypes created with a modeling software package (such as, for instance, MATLAB®) can be connected to the process as long as they support the OPC communication protocol.

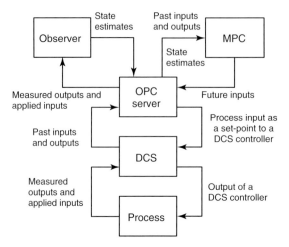

Figure 16.5 Structure of the MPC system.

16.10
MPC of Crystallization Processes

A few applications of MPC to continuous crystallization were reported in the literature. The discussed applications concern both SISO and MIMO systems.

Eek (1995) implemented and tested an MPC controller for a 970 l draft-tube-baffle (DTB) crystallizer operated isothermally in evaporative mode to crystallize ammonium sulfate from its aqueous solution. In this work, the closed-loop response of an MPC-based control system was compared to that of a (multiple-loop) PID control system.

Three control configurations for a continuous crystallization process with a general form as shown in Figure 16.6 were studied. The goal of the study was to evaluate the performance of different control schemes with respect to disturbance rejection, set-point tracking, and stabilization of the process. Injection of water and blockage of product removal was considered as the process disturbances.

The first configuration was a SISO system with the fines removal rate as an input and a variable related to the CSD as an output. For this system, both control schemes were evaluated experimentally. It was found that a SISO proportional-integral (PI) controller performed as good as an MPC controller with a state estimator. It was also shown that the MPC controller without the state estimator performed much worse than the PI controller.

Clear advantages of MPC were revealed during the experimental tests and simulations of control configurations for MIMO systems. First, a two-input–two-output system was considered. The SISO system described above was augmented with one manipulated variable (heat input to the crystallizer) and one controlled variable (the crystallizer magma density). Design of a MIMO MPC controller was based on two different linear models. The first model was derived by linearization of a nonlinear first-principles PBE-based model with the subsequent model reduction of the high-order linear model. The second model was identified from the closed-loop response of the process (see Eek (1995) for details).

As the two input–output pairs were uncorrelated, the process responses of all control configurations were similar. However, the process disturbances were better suppressed by the MPC configurations. The MPC controller built on the identified model yielded slightly better results. This could be attributed to the identified model being more accurate than that derived from a first-principles model.

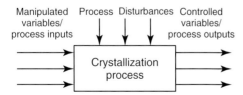

Figure 16.6 Schematic representation of a crystallization process.

In order to compare the performance of the MIMO MPC and PI controllers, a simulation study with a three-input–three-output system was carried out. The additional input–output pair was the product flow – the supersaturation. It was demonstrated that the MPC controller was superior to the multiloop PI controller as the former could take explicitly into account the interaction between the heat input and the supersaturation.

Rohani *et al.* (1999a, b) extended the work of Eek (1995). They compared the performance of two MPC configurations that were based on a linear model on the one hand and nonlinear models on the other hand. The linear model was obtained during a three-step identification procedure. The model identification was similar to the one used by Jager *et al.* (1992) to identify a linear SS model for the same crystallization system that was considered by Eek (1995). Nonlinear neural network models were used as an alternative to the linear model.

The studied process was a continuous cooling crystallization of KCl. The crystallization process was a three-input–three-output system. The inputs were the fines removal rate, the clear liquor advance rate, and the crystallizer temperature. The outputs were a variable related to supersaturation (for control of the product purity), a variable related to the fines suspension density (for control of the CSD), and the magma density.

The conclusions of the study by Rohani *et al.* (1999a, b) were that an offset-free control of the outputs could not be achieved using the linear model-based MPC. This shortcoming was attributed to model inaccuracies. The MPC controller based on the nonlinear models was better in tracking the set-point changes even under a simulated mismatch between the process and the model.

16.11
Delta-Mode MPC

MPC of a transient continuous process can be formulated to follow an optimal trajectory computed by a dynamic optimizer. The dynamic optimizer can be easily added to the advanced control architecture (Figure 16.5). The optimizer computes optimal trajectories for input and output variables using an objective function that can take into account some economic measures such as operating costs and transition time. An optimal transition from one steady state to another upon a set-point change is an example of the transient operation of a continuous process.

A block diagram of the delta-mode MPC with the dynamic optimizer is given in Figure 16.7. The dynamic optimizer can be a real-time optimization application communicating with the MPC controller and the rest of the components in Figure 16.5 via the OPC server. The optimal trajectories of the process inputs and outputs can also be obtained during off-line optimization. In that case, they are made available to the MPC controller through the OPC server.

The predictive controller penalizes the differences between the actual input and output trajectories and the reference trajectories by minimizing a cost function

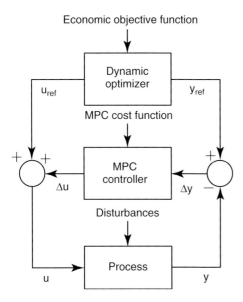

Figure 16.7 Delta-mode MPC with a dynamic optimizer.

similar to the one by Equation (16.3). In order to force the actual input signal to follow the reference input signal, an additional term can be added to the cost function given by Equation (16.3)

$$\sum_{j=1}^{N_C} \left(\mathbf{u}\left(k+j-1\right) - \mathbf{u}_{\text{ref}}\left(k+j-1\right)\right)^{\mathrm{T}} \mathbf{S} \left(\mathbf{u}\left(k+j-1\right) - \mathbf{u}_{\text{ref}}\left(k+j-1\right)\right) \quad (16.8)$$

where **S** is a positive definite diagonal weighting matrix.

Batch operation is similar to a transient operation of continuous crystallization processes. MPC can also be used for control of batch processes in the way described above. An example of a batch process control that was implemented according to the presented architecture (Figure 16.5) and used a delta-mode model predictive controller is given in Chapter 14.

16.12
Conclusions and Perspectives

Model predictive control is a multivariable control technique suitable for control of dynamic MIMO systems with complex interactions between the inputs and the outputs. A distinct advantage of MPC over other control techniques is that constraints on the process variables are directly included in the controller formulation. As an optimal process operation is often an operation close to the constraints on the process variables, MPC can ensure that the constraints are not violated during operation close to the limits. Moreover, MPC can achieve a more optimal process operation close to an economically justifiable limit by first reducing the variance of a controlled process variable and then pushing that variable closer to the optimal constraint limit. MPC is flexible in operation with the performance criteria and the process conditions changing rapidly in response to the dynamic market demands.

Despite successful application of MPC in the chemical industry, applications for control of crystallization processes are scarce. A highly nonlinear behavior of crystallization processes imposes demanding requirements on the process model. A single linear model used in the linear MPC algorithms is often not accurate enough in predicting the transient process behavior. It is not until recently that predictive crystallization models are becoming available. This explains the rising number of publications reporting computer simulations of predictive control of crystallization processes.

Nonlinear MPC algorithms are also being developed to improve the MPC performance for such nonlinear process systems as crystallization. The developed nonlinear control algorithms are either extensions of the linear MPC that utilize linear time-varying models or control techniques that make a direct use of nonlinear process models.

As for implementation of MPC, it is still limited by availability of reliable measurements of the solute concentration and the CSD. Most of the designs of the industrial crystallizers also lack effective actuators that are able to satisfy the desired control objectives such as, for instance, producing a product with a defined CSD.

References

Agachi, P., Nagy, Z., Cristea, M., and Imre-Lucaci, A. (2006) *Model Based Control. Case Studies in Process Engineering*, Wiley-VCH Verlag GmbH, Weinheim.

Camacho, E. and Bordons, C. (1999) *Model Predictive Control*, Springer, London.

Dochain, D. (2003) State and parameter estimation in chemical and biochemical process: a tutorial. *J. Process Control*, **13**, 801–818.

Eek, R. (1995) Control and dynamic modelling of industrial suspension crystallizers. PhD thesis. Delft University of Technology, The Netherlands. http://www.library.tudelft.nl.

Froisy, J. (2006) Model predictive control – building a bridge between theory and practice. *Comput. Chem. Eng.*, **30**, 1426–1435.

Jager, J., Kramer, H., De Jong, E., De Wolf, S., Bosgra, O., Boxman, A., Merkus, H., and Scarlett, B. (1992) Control of industrial crystallizers. *Powder Technol.*, **69**, 11–20.

Kalbasenka, A.N. (2009) Model-Based control of industrial batch crystallizers: experiments on enhanced controllability by seeding actuation. PhD thesis. Delft University of Technology, The Netherlands. ISBN: 978-90-6464-361-3. http://www.dcsc.tudelft.nl/Research/PhDtheses/.

Kalbasenka, A., Landlust, J., Huesman, A., and Kramer, H. (2006) Application of observation techniques in a model predictive control framework of fed-batch crystallization of ammonium sulphate. AIChE Spring Meeting Conference Proceedings, Vol. 2: 5th World Congress on Particle Technology, April 23–27, Orlando, FL.

Ljung, L. (1999) *System Identification. Theory for the User*, 2nd edn, Prentice-Hall, Upper Saddle River, NJ.

Maciejowski, J. (2002) *Predictive Control With Constraints*, Prentice-Hall, Harlow.

Muske, K. and Rawlings, J. (1993) Model predictive control with linear models. *AIChE J.*, **39** (2), 262–287.

Qin, S. (2006) An overview of subspace identification. *Comput. Chem. Eng.*, **30**, 1502–1513.

Rohani, S., Haeri, M., and Wood, H. (1999a) Modeling and control of a continuous crystallization process: Part 1. Linear and non-linear modelling. *Comput. Chem. Eng.*, **23**, 263–277.

Rohani, S., Haeri, M., and Wood, H. (1999b) Modeling and control of a continuous crystallization process: Part 2. Model predictive control. *Comput. Chem. Eng.*, **23**, 279–286.

Rossiter, J.A. (2003) *Model-Based Predictive Control: A Practical Approach*, Control Series, (ed. R. H. Bishop), CRC Press, Boca Raton, FL.

Tadayyon, A. and Rohani, S. (2001) Extended Kalman filter-based nonlinear model predictive control of a continuous KCl–NaCl crystallizer. *Can. J. Chem. Eng.*, **79**, 255–262.

Van Brempt, W., Backx, T., Ludlage, J., Van Overschee, P., De Moor, B., and Tousain, R. (2001) A high performance model predictive controller: application on a polyethylene gas phase reactor. *Control Eng. Pract.*, **9**, 829–835.

17
Industrial Crystallizers Design and Control
Franco Paroli

17.1
Introduction

Continuous crystallizers are used in the industry for the production of chemical commodities as well as for the recovery of valuable products from waste streams. Production capacities can vary from 12 to 1200 metric tons per day.

According to specific process requirements or production economics, crystallizers can be arranged in several ways, product-wise or energy-wise, reaching high levels of plant complexity: single or multiple stages, single or multiple effects, with or without thermal or mechanical vapor recompression.

The operation of a crystallization plant should ensure:

- the production capacity and yield;
- stable and trouble-free operation;
- the required product purity; and
- optimal crystal size distribution (CSD).

To reach such goals, the designer must focus on four areas, and all of them are of equal importance:

- process flow scheme;
- process conditions;
- mechanical design of the crystallization equipment; and
- design of the control system and selection of the proper instruments.

The design of an effective control system is very important. Continuous crystallizers should ensure constant growing conditions for the crystalline product but, due to the large dimensions of this equipment (up to $400 \, m^3$ for a single vessel), achieving uniform distribution of temperature, solute concentration, and crystal suspension is not always possible.

Specific values of operating parameters in continuous industrial crystallizers, such as supersaturation and residence time of crystals, are a matter of the appropriate design of the system; they should be maintained at steady-state operation by monitoring and control of a series of process variables. Some of these

Industrial Crystallization Process Monitoring and Control, First Edition. Edited by
Angelo Chianese and Herman J. M. Kramer.
© 2012 Wiley-VCH Verlag GmbH & Co. KGaA. Published 2012 by Wiley-VCH Verlag GmbH & Co. KGaA.

are traditional chemical plant process variables (such as pressure, temperature, density, level, flow rate, and pH) and their feedback control is required to achieve

- stable and smooth operation;
- fixed production capacity;
- high yield;
- low-energy consumption; and
- low fouling/encrustation levels.

Other control variables are more specific for crystallization processes (such as crystal surface, nucleation rate, and mixing input power) and their control is required to achieve targets of

- product quality;
- crystal size; and
- on-stream operating rates.

Unfortunately, in industrial systems, feedback control of these last variables is not commonly practiced, mostly due to the lack of reliable instruments at affordable cost. Therefore, the applied control of industrial crystallizers is usually limited to the one of traditional modern chemical plants, based on the feedback control of single operating variables.

In this chapter, firstly process and instrumentation schemes of the most commonly used types of crystallizers (forced circulation (FC), draft-tube-baffle (DTB), and Oslo growth-type) are presented, then the used sensors and control devices in the industrial practice are shown.

17.2
Forced Circulation Crystallizer

The FC crystallizer is the practical representation of the mixed suspension mixed product removal (MSMPR) crystallizer. It is composed of a crystallization vessel and an external circulation loop, which includes a propeller pump to provide the mixing requirement, and, if necessary, with a shell and tube heat exchanger to add or remove heat to the crystallizer. The mixing, provided by the external circulator, in combination with the requirements of thermal exchange, provides the operating supersaturation for the crystallizer (Mersmann, 2001).

Due to its simplicity of design and operation, it is the "workhorse" of the chemical industry and is well suited to be used in single as well as multiple effect arrangements, without re-use of the vapors produced, or with reuse either by thermocompressors (steam driven ejectors) or by mechanical compressors.

The FC is usually employed in evaporative crystallization systems with a solubility trend with temperature relatively "flat" (such as for sodium chloride) or inverse (such as for sodium sulfate) solubility, or with viscous process liquor. It is also very useful for applications where scaling is a major problem, such as zero-liquid-discharge crystallizers.

The typical arrangement of a single effect steam heated FC crystallizer, operating under vacuum, is shown in Figure 17.1. Since this system is characterized by evaporation, the control philosophy is to ensure the material balance: feed is admitted to replace the water removed by evaporation and the precipitated crystals removed by dewatering, by controlling the crystallizer level. The required capacity (expressed as a given feed flow) is obtained by controlling the heating steam flow (energy input to the system), which will determine the rate of the evaporated water and consequently of the crystallization. The steam will be de-superheated by condensate injection before entering the heater. The crystallizer operates at constant suspension density (the control of suspension density determines both residence time and total crystal surface area). Extraction of the product crystals is achieved by the slurry density control, which has as manipulating variable the flow rate of the slurry stream sent to centrifugation. In the example, such slurry stream is withdrawn from a loop kept at constant flow in order to minimize plugging of the line. Feed is mixed with the filtrate mother liquor, to dissolve small crystals that have passed through the centrifuge screen. The combined fluid is then sent to the crystallizer.

The operating temperature is maintained by controlling the pressure in the vapor space of the crystallizer vessel, by injecting ambient air at the inlet to the vacuum pump.

Indication of the tube-side temperature difference across the heat exchanger is very important in determining the operating supersaturation and heat transfer conditions.

Periodic washing (with timer controls on frequency and duration) of the demister and the vessel wall above the boiling liquid level complete the control system.

The process flow diagram of the thermal vapor recompression crystallizer shown in Figure 17.2 is very similar to the previous one.

The main difference is the crystal extraction system: in this case, the suspension density inside the crystallizer is lower than the value required by the centrifuge (in this case a pusher-type centrifuge, which requires solids content in its feed >40%wt, too high a value for an FC crystallizer), so a thickening hydrocyclone is installed ahead of the centrifuge. Slurry extraction rate (and thus suspension density) is controlled by varying the speed of the extraction pump. In using this method of discharge control, the pump speed range must be such that the pipe velocity is always high enough to avoid sedimentation and assure proper hydrocyclone operation. In the case of operational upsets, low density, or centrifuge failure, slurry is recycled back to the crystallizer by a three-way valve. Capacity of the plant is determined by the amount of motive steam to the thermocompressor.

The plant shown in Figure 17.3 is similar to the thermal recompression system, but in this case the process vapors are compressed by a mechanical compressor.

Capacity of the plant is determined by the compressor speed (if a positive displacement volumetric unit or high-speed radial fan) or by inlet guide vanes orientation (if a centrifugal unit). The operating pressure (and thus the temperature) of the crystallizer is kept constant by adding make-up steam (in the case of energy deficiency) or by venting heating vapors (in the case of excess of energy).

17 Industrial Crystallizers Design and Control

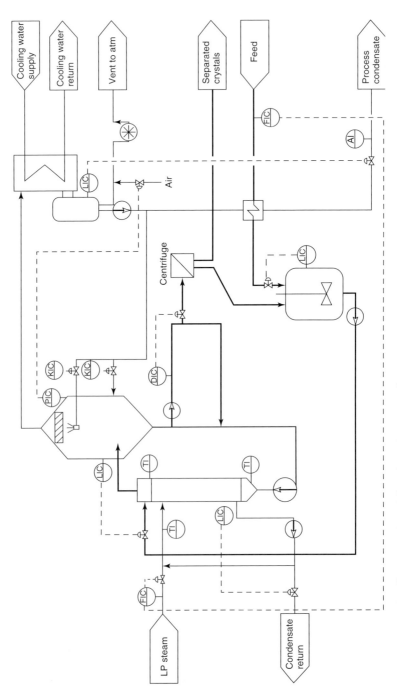

Figure 17.1 Single effect, steam heated forced circulation crystallizer.

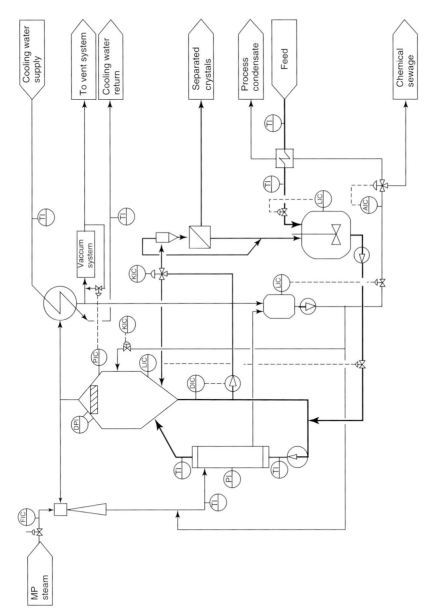

Figure 17.2 Single effect forced circulation crystallizer, with thermal vapor recompression.

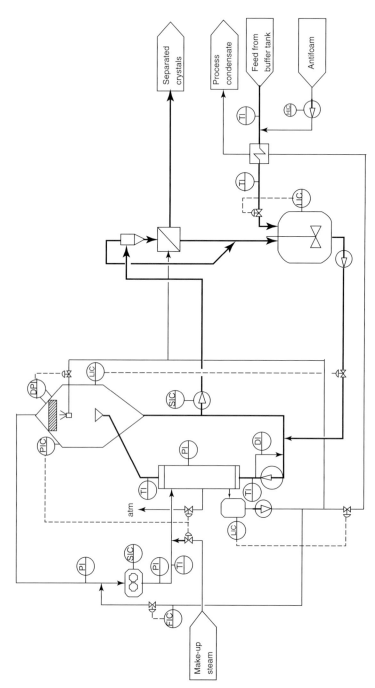

Figure 17.3 Single effect forced circulation crystallizer, with mechanical vapor recompression.

Figure 17.4 illustrates a steam-heated triple effect FC crystallization system of the type commonly used to produce sodium chloride by recrystallization of crude salt of sea or rock origin.

The goal of such plants is to produce a pure product while minimizing scaling of the system, which is usually due to both salting (by the product itself) and fouling (by foreign substances). Special attention is given to the design of the crystallization loop (the "reverse-Messo type" is shown here, which minimizes formation of encrustations on the walls), and to the counter-current fluidized column ("elutriation leg") that removes foreign particles such as gypsum crystals from the discharge slurry and provides narrow-size distribution, prewashed crystal feed to dewatering centrifuge.

17.3
Draft-Tube-Baffle Crystallizer

The DTB crystallizer is an equipment designed to produce coarse crystals with narrow-size distribution. It consists of a vessel where growing crystals suspended in their mother liquor are gently agitated by an up-pumping circulator (propeller type) located inside a draft tube, and surrounded by an annular baffle from which a stream of mother liquor with fine crystals in suspension is extracted. This stream (baffle overflow) may be sent to the next crystallizer (in the case of a multistage vacuum cooling line) or recycled through a heat exchanger in case of evaporative systems. Re-dissolution of such fines by heating or dilution of the baffle overflow allows control of the crystal population and production of large-size particles. The product crystals are extracted as slurry and sent to dewatering equipment, and the filtrate is recycled back to the crystallizer (Bennett, 1993).

Key control points for the DTB crystallizer are the mass and size of fine crystals to be redissolved. This is done by varying the baffle overflow (higher upward flow contains higher mass and larger particle sizes). The actual suspended solids content and size distribution is usually measured by manual sampling, although there exist several devices that can do this in-line, such as laser backscattering devices, ultrasonic particle detectors, or turbidity meters.

Figure 17.5 illustrates the typical arrangement of a single effect DTB evaporative crystallizer, operating under vacuum with thermal recompression of the process vapors.

The illustration shows a system typically used to produce ammonium sulfate from waste streams, such as in the methyl-methacrylate industry. The DTB shown here is an evaporative crystallizer, and its capacity is determined by the amount of motive steam to the thermocompressor. The crystallizer is fed by level control. Suspension density inside the body is kept constant by adjusting the extraction rate of slurry to the centrifuge; filtrate is recycled to the discharge pump suction, so that the slurry piping always operates with sufficient pipe velocity and settling of crystals in the line is avoided. A particle detection/measurement device is installed in-line in the baffle overflow pipe to measure the amount of the crystals carried

210 | 17 Industrial Crystallizers Design and Control

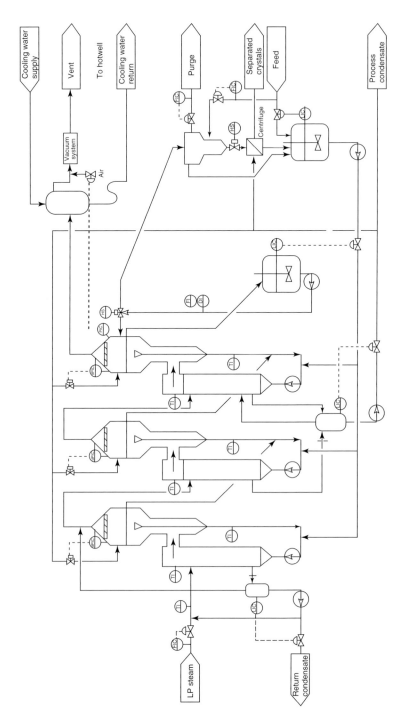

Figure 17.4 Triple effect FC-type evaporative crystallizer with washing thickener.

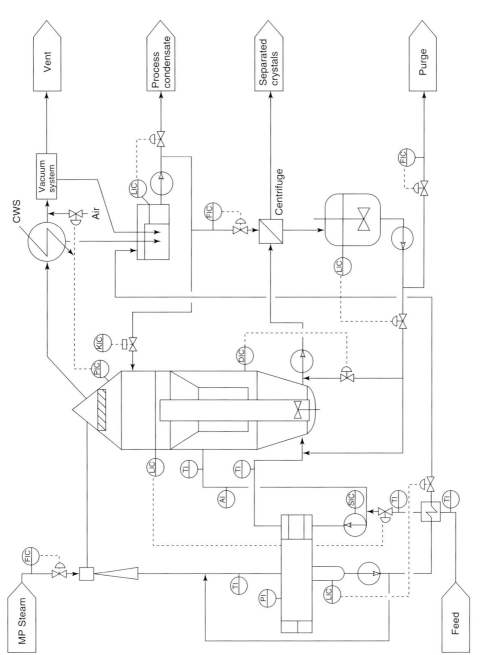

Figure 17.5 Single effect DTB evaporative crystallizer, with thermal vapor recompression.

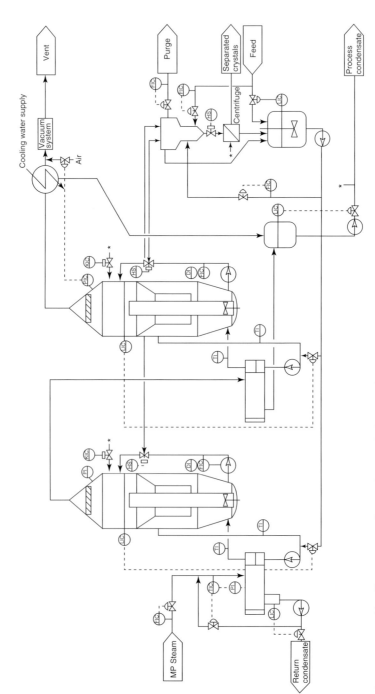

Figure 17.6 Double effect DTB crystallizer with external elutriation leg.

on in that stream. The baffle overflow circulation pump is equipped with variable speed drive (under manual control!), in order to adjust the baffle overflow rate in the crystallizer annulus without upsetting the thermal exchange.

Impurity build-up in the system is avoided by purging under flow control.

The system shown in Figure 17.6 is a double effect DTB crystallizer designed to produce uniform and coarse ammonium sulfate crystals from waste streams containing organic impurities (soluble or as a separate oily phase) such as in the caprolactam industry.

The system is equipped with an external elutriation leg, allowing further crystal classification and removing undesirable fines. These fines are recycled to the DTB. Another function of the elutriation leg is to remove soluble impurities from the crystal surfaces, by replacing the concentrated liquor that surrounds the crystals with fresh feed, and to separate oily phases by skimming. Performance of the leg is controlled by adjusting the various upward flows.

By its very nature, the DTB crystallizer produces a cyclic CSD (Van Esch et al., 2008). Sometimes, as in the case shown above, such effect may be dampened by the fact that two crystallizers have different cycling periods; recycling of fines from the elutriation leg via the feed tank can provide a stabilizing effect on the CSD. The degree and severity of cycling is dependent on the crystal size produced and other factors. For many applications, the cycling effect on the product quality can be negligible.

Figure 17.7 shows a reactive DTB crystallizer: This type is well suited for the production of crystalline materials by direct reaction of substances such as ammonia (liquid or gaseous) and sulfuric acid, taking advantage of the heat of reaction for evaporation purposes, reducing energy consumption, and avoiding upstream reactors.

In the reactive DTB, the reaction is carried on in a suspension of growing crystals and the heat of reaction is removed by vaporizing water that, in some cases, can be recycled into the baffle overflow, providing useful dissolution effect for fines control. Injection of the reactants can be either located in the active volume of the crystallizer vessel, or in the external circulation loop (when the production of coarse crystals is required). In both cases, good mixing should be ensured at the feed point to avoid temperature and supersaturation peaks.

The reaction rate is primarily controlled by flow control of the reactants, and adjusted by a pH cascade control if quality of reactants is not constant. Special attention must be given to proper gaseous reactant injection; in some cases, in order to keep constant high velocity across the spargers the gaseous reactant is mixed with steam prior to its injection.

17.3.1
"Oslo" Growth-Type Crystallizer

The Oslo crystallizer operating principle is to allow the crystals to grow without mechanical attrition, in a fluidized bed, with the aim to produce a coarse product. Typically, this crystallizer consists of two sections: a fluidized bed in the conical

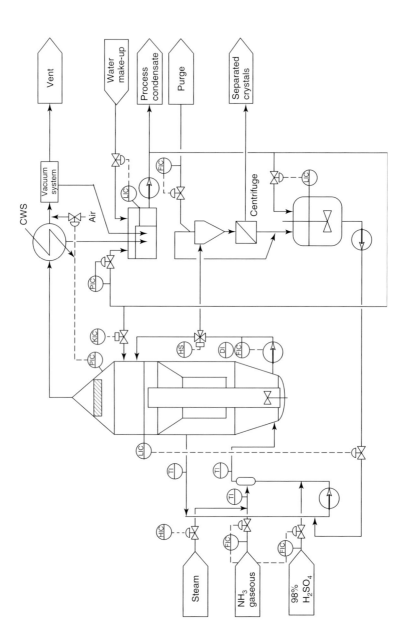

Figure 17.7 Reactive DTB crystallizer.

part of a clarifier (bottom section), and a volume above the fluidized bed relatively clear of suspended particles (upper section). Overflow from the clarifier is pumped through a heat exchanger, to absorb the energy the unit needs to generate the necessary evaporation, and then it is returned in the crystallizer's vapor/liquid separator. After release of the evaporation, the now supersaturated liquid flows through a downcomer, to become the fluidizing solution for the crystal bed. This system configuration is typical of double draw off systems, which provide different residence times: higher for solid with respect to that of mother liquor (Wohlk and Hofmann, 1987).

Since the circulation rate is set by the needs of internal agitation, crystal bed fluidization, supersaturation, and thermal exchange duty, it must be checked carefully to fulfill all these process requirements. The low circulation rate will result in low secondary nucleation, leading to bigger crystals, while too low circulation will push the system toward primary nucleation, increasing supersaturation that may result in encrustations. At the same time, under a threshold value there will be the risk to lose suspension all together. On the contrary, an excessive circulation rate can upset the fluidized bed and carry over crystals in the external loop, leading the system to operate as a FC crystallizer.

While Oslo crystallizers can produce very large crystals, they often suffer from severe encrustation formation, because the vapor/liquid separator and downcomer operate with highly supersaturated liquid.

Oslo crystallizers are not very popular any more, but still remain the best choice for many organic crystals like amino-acids, substances that are soft and/or fragile (mechanical attrition would easily reduce their size) and have low growth rates.

Figure 17.8 illustrates a double stage Oslo flash cooling system; both crystallizers are operated at controlled suspension density, by separate discharge of mother liquor (by level control), and crystals suspension (by slurry density control). To lessen encrustations inside and at the bottom of the downcomer that reduces circulation (this has a snow-ball effect on circulation with increasing scale formation) both crystallizers of the unit shown are equipped with a variable-speed circulator, controlled by the temperature difference upstream and downstream of the feed point on the circulation loop (the set-point value is proportional to the circulation flow).

Another traditional field of application of Oslo crystallizers is in the caprolactam industry, where reactive crystallization is often used to neutralize the lactam ester and thus to separate lactam oil while precipitating ammonium sulfate. This process takes advantage of the heat of reaction to evaporate water contained in the oximation waste stream (saturated with $(NH_4)_2SO_4$). In such an application, shown in Figure 17.9, the geometric characteristics of the Oslo allow a very efficient separation, with very low salt contamination, of the oily phase.

In the unit shown earlier, lactam ester (the main stream to be treated) is fed on flow control, with neutralizing ammonia flow controlled by pH, while the ammonium sulfate solution from oximation is fed on the crystallizer level control, being a function of the available evaporation capacity.

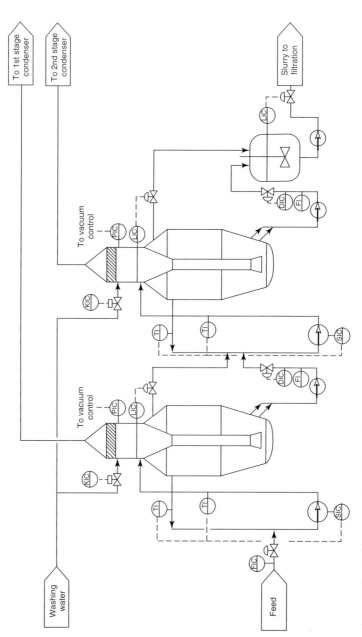

Figure 17.8 Double-stage Oslo cooling crystallization system.

17.3 Draft-Tube-Baffle Crystallizer

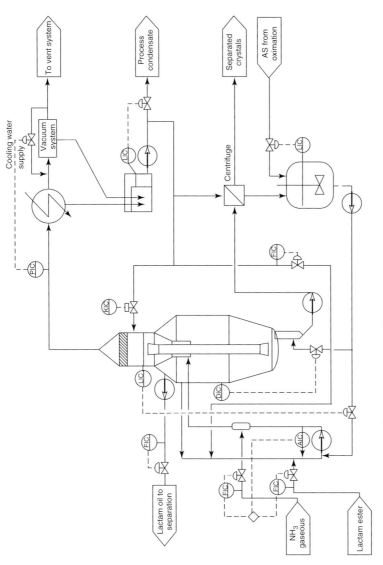

Figure 17.9 Single-stage Oslo-type reactive crystallizer.

Suspension density inside the body is kept constant by adjusting the extraction rate of slurry to the centrifuge.

In such a unit, the circulation flows inward from the wall of the vapor/liquid separator to the separator center, where it descends through the downcomer. Such arrangement has the least degree of supersaturation at the wall, thus minimizing encrustations there; moreover, together with the use of a discharge elutriation leg with lump catcher, it allows the crystallizer to achieve long operating cycles, similar to the ones of DTBs.

17.4
Process Variables in Crystallizer Operation

Crystallizer operation can be very customized to reflect the needs of very special operation procedures and conditions. However, there are certain variables that are common to most crystallizers.

17.4.1
Continuous Operation

The following variables are usually controlled by methods discussed in more detail in the above-described crystallizers.

- Operating temperature (if surface-cooled crystallizer).
- Crystallizer level (gas–liquid interface).
- Absolute pressure (if evaporative/vacuum cooled crystallizers), which determines the operating temperature.
- Slurry density (via mother liquor recycle).
- Energy input or removal rate (steam to heater, cooling medium rate, or temperature to surface cooler).

The following parameters are usually not controlled by the crystallizer operation, but are significant in proper crystallizer operation.

- *Feed rate*: A crystallizer is usually expected to handle the feed rate coming from upstream unit operations. In a typical control scheme, the feed rate sets the energy input/removal rate for the crystallizer.
- *Feed temperature*: The same comments as for the feed rate apply.
- *Crystal size*: A very important aspect of the crystallizer operation, it is usually not controlled on real-time basis due to severe influence that such control might have on other aspects of the crystallizer operation, such as production and centrifuge capacity, and so on. Step-wise changes are made, sometimes, to bring about a long-term change in the crystal size, such as speed up or slowdown of the crystallizer mixer, fines destruction capability, and so on.
- *Agitator or pump speed*: This parameter is rarely controlled in continuous mode; since it influences the crystal size, its changes are made as a result of corrective actions and are usually permanent.

- *Crystal residence time*: This is actually controlled via the slurry density, which is a continuous control value; large changes to the target value for this parameter can mean serious changes to equipment downstream of the crystallizer (going from 15 to 30%wt crystals in the crystallizer can mean changes in slurry pipe size; if the change is in the opposite direction, the slurry pump may be undersized).

17.4.2
Batch Operations

Batch crystallization is relatively simple to operate and control, and much more adaptable to fully automatic operation. Fill and drain operations, for example, are usually automated, and such parameters as slurry density and mixer speed can be adjusted either before each batch or during a batch, as a function of operation time. Typical batch operation parameters that are controlled are as follows.

- *Rate of temperature change*: It is controlled by heat removal rate for surface-cooled crystallizers (usually set by the degree of supersaturation desired for the optimum crystal growth rate).
- *Rate of pressure change*: As with continuous evaporative crystallizers, batch evaporative or vacuum cooled crystallizers have their temperature controlled by controlling the crystallizer pressure (with above-discussed methods). This controls the supersaturation in the crystallizer.
- *Agitator or pump speed*: This may be changed over the batch duration, to reflect the changing mixing regime in the crystallizer. Optimization of mixing can result in improved crystal size, by reducing secondary nucleation. It can be done by a timer, or by sensing a related parameter, such as magma density.
- *Batch time*: This parameter is usually a main boundary for the crystallizer design, and cannot be changed easily, because the peripheral equipment is usually sized according to the design batch time. Minor adjustments can be possible, and are used to make small improvements to the crystal size (for example, by allowing a short "hold" time at the batch end, in order to reach the wished final de-supersaturation).

17.5
Sensors

The sensors used in industrial scale crystallization plants are the same as those used in traditional chemical plants, but specially arranged to avoid fouling (sedimentation of the solids in suspension, encrustations) during operation. Operation under boiling conditions, under vacuum or internal pressure is also normal.

The sensors that are most affected by slurry service, and require special features and installation arrangements are the following.

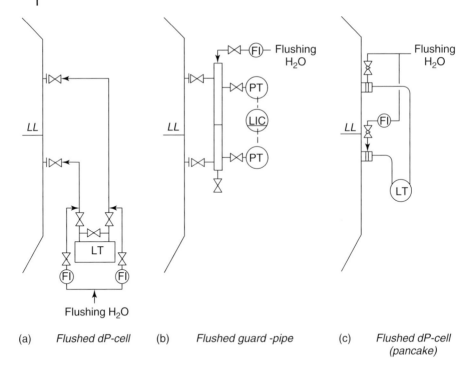

(a) Flushed dP-cell (b) Flushed guard-pipe (c) Flushed dP-cell (pancake)

Figure 17.10 Level measurement arrangement in crystallizers: (a) flushed DP-cell; (b) flushed guard-pipe; (c) flushed DP-cell (pancake).

17.5.1
Level

A reliable measurement of the level inside the crystallizer at the vapor–liquid interface is one of the most important wishes of the operators; apart from radioactive devices (an application that has safety constraints) and radar devices (a rather recent application) the most popular option remains the use of differential pressure (DP) transmitters. Figure 17.10 shows three proven arrangements fitted with a water purge system commonly used in crystallizer service.

17.5.2
Density

Similar considerations can be applied to density measurement, devoted to measure real density of the liquor or suspension density (as % wt of suspended crystals) of a slurry. The two most reliable measurement devices are: (i) the weight measurement instruments consisting of a flushed differential pressure transmitter installed on a circulation pipe (see Figure 17.11a) or on the crystallizer vessel and (ii) a mass flow ("Coriolis") transmitter, which can be installed on a bypass of the circulation

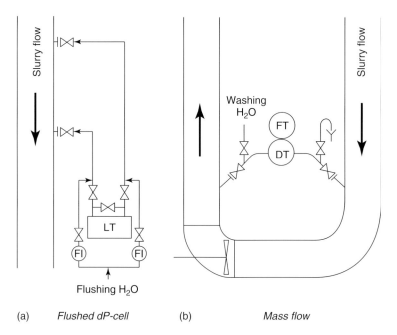

Figure 17.11 Density measurement instruments: (a) flushed DP-cell; (b) mass flow.

pump (see Figure 17.11b) or on a slurry line. In the case of the FC crystallizer, the installation of a mass flow on a bypass of the circulation pump has, apart from the possibility of washing the device in operation, the advantage to give indication on the variation of the circulation flow rate.

17.5.3
Crystal Size

Because of the high initial cost of the several in-line particle characterization devices on the market (light- and sound-based), and some apprehension about the repeatability of data produced by these instruments, most continuous installations use, instead, manual monitoring of the crystal size in the crystallizer. However, as personnel costs become a target for cost saving in many plants, and to avoid, for many crystallizer operations, potential hazards involving personnel collecting and/or handling the sample, in-line particle size sensors are becoming more popular. In the pharmaceutical industry, these devices have been used extensively for research purposes, and for batch operations, usually to provide an early warning of poor batch results. The use of any of these devices in a continuous crystallizer should be made with the same features as the sensors discussed in this chapter: the ability to wash the sensor parts that are exposed to the process (and may suffer from scale formation or plugging) is very important. The location of these sensors is also a concern: their accuracy may be affected by local values of slurry density or particle size distribution, as well as the flow lines around the sensor, which may selectively remove some particle sizes from the sample exposed to the sensor.

17.6
Control Devices

The most common operating device is still the control valve. When operating on suspensions, special types of valves must be used in order to avoid/minimize the risk of plugging. Some of the control valves most frequently used in such service are listed below, with indication of their main features.

- *Plug valve*: It is one of the oldest valve designs, originally sold as quick opening ("quarter turn"). Special plug designs allow good flow control in slurry service (see Figure 17.12). Its advantages are the full port opening (flow not constricted), no internal cavities, and self-cleaning: at each rotation of the valve, encrustation that may have developed on the plug seal are broken out and flushed away. Due to their conceptual design, plug valves can be fabricated in solid material or sleeved with fluoro-polymers.
- *Eccentric disk rotary control valve*: It offers very effective throttling control, providing linear flow characteristic through 90° of disk rotation (see Figure 17.13). Eccentric mounting of the disk plug pulls it away from seal, afterward it begins to open, minimizing seal wear. The design is usually very compact. Various plug designs (ball-, V-shaped, etc.) are available for different services.
- *V-notch ball control valve*: The construction is similar to a conventional ball valve, but with a contoured V-notch in the ball producing an equal-percentage flow characteristics (see Figure 17.14). Such valves operate well over a wide range of settings, control, and shut-off capability, making them well suited for slurry flow control (even in erosive/abrasive service). The ball remains in contact with

Figure 17.12 Plug valve.

Figure 17.13 Eccentric plug control valve.

Figure 17.14 V-notch ball control valve.

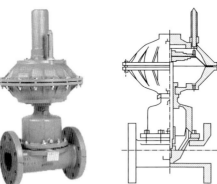

Figure 17.15 Diaphragm valve.

the seal during rotation, which produces a shearing effect as the ball closes and minimizes clogging.
- *Diaphragm valve*: It is available as weir- or straight-through type, this last being ideal for slurry service. Its features do not allow solid particles to be trapped in cavities or crevices that may obstruct the operation (see Figure 17.15). This valve provides linear operation with high coefficient of variations (CVs) and very low pressure drops. It is usually available in the cast iron body with internal plastic or rubber lining. It is an efficient and inexpensive valve but requires frequent service (replacement of the diaphragm due to wear).
- *Pinch valve*: It has been developed for use in thick slurries. It consists of a metallic body and a natural or synthetic rubber sleeve, which is resistant to erosion

Figure 17.16 Pinch control valve.

(see Figure 17.16). Closing is achieved by squeezing the sleeve by a pinch mechanism, operated mechanically or pneumatically. Since the sleeve is the only wetted part, no seals or stuffing boxes are necessary.

In many cases, it can be useful to avoid the use of control valves on slurry service, by using variable speed pumps, that can be centrifugal (open or backward impeller type) or volumetric (such as the progressive cavity type), operated by frequency controlling devices or similar equipment.

References

Bennett, R.C. (1993) Crystallizer selection and design, in *Handbook of Industrial Crystallization*, Butterworth-Heinemann, Oxford, 103–130.

Mersmann, A. (2001) *Crystallization Technology Handbook*, Marcel Dekker, New York

Van Esch, J., Fakatselis, T.E., Paroli, F., Scholz, R., and Hofmann, G. (2008) Ammonium sulphate crystallization – state of the art and trends. Proceedings of the 17th International Symposium on Industrial Crystallization.

Wohlk, W. and Hofmann, G. (1987) Types of crystallizers. *Int. Chem. Eng.*, **27** (2), 197–204.

Index

a

absorbance   29, 31–33
actuators and basic recipe control
    108–109
– seeding as   111–112
– – initial supersaturation   112
– – seed addition methods   114
– – seed mass   112–113
– – seed quality and preparation procedure
        113–114
– – seed size and size distribution   113
advanced model-based recipe control
    161–163
– MPC for batch crystallization   168–172
– online dynamic optimization   163–168
advanced recipe control   139
– dynamic optimization   153–154
– experimental validation results   155–157
– implementation   150
– incentives and strategy of   139–140
– mixing conditions   145–150
– modeling for optimization, prediction, and
        control   141–143
– model validation   143
– objectives   150–151
– process description and modeling
        151–153
– supersaturation generation rate   144–145
anomalous diffraction   12
Attenuated total reflectance Fourier transform
    infrared (ATR-FIR) spectroscopy   81, 107,
    119
– calibration   82, 83
– co-crystal formation   84–86
– crystal growth rates   86
– crystallization monitoring and control
        89
– impurity monitoring   89
– polymorph transformation   87–88
– solubility measurement   86
– speciation monitoring   84
attenuation   51
axis length distribution (ALD)   39

b

basic recipe control   105
– actuators   108–109
– crystallization mechanisms   106
– crystal number, size, distribution, and
    morphology   107–108
– fines removal and dissolution   118–119
– implementation   119–122
– incentives for   105–106
– mixing and suspension of solids   116–118
– seeding as process actuator   111–112
– – initial supersaturation   112
– – seed addition methods   114
– – seed mass   112–113
– – seed quality and preparation procedure
        113–114
– – seed size and size distribution   113
– sensors   106–107
– solute concentration measurement   107
– strategy   109–110
– – obtaining   110–111
– – scaling   111
– supersaturation generation rate   114–116
batch crystallization   105, 115, 117, 119, 120,
    121, 122, 142, 145, *176*
– MPC for   168–172
– seeding technique in   127
– – control   131–137
– – main process parameters   128–130
– – principles and phenomena   127–128
Boltzmann's law   93
box area   38

Industrial Crystallization Process Monitoring and Control, First Edition. Edited by
Angelo Chianese and Herman J. M. Kramer.
© 2012 Wiley-VCH Verlag GmbH & Co. KGaA. Published 2012 by Wiley-VCH Verlag GmbH & Co. KGaA.

c

chord length distribution (CLD)   21, 23, 26, 37
closed-loop optimal control system   161, 162, 163, 164, 166, 171
– online   163
– open-loop profiles vs.   167
coefficient of variation (CV)   4
computational fluid dynamics (CFD)   116
continuous crystallization   197, 199
control system   203, 205
critical angle refractometer   72, 76
crystallization   175–176. See also individual entries
crystallizers   25, 26
crystal size distribution (CSD)   1–6, 17, 29, 31, 33, 35, 36, 37, 46, 78, 105, 111, 113, 114, 116–118, 119, 140, 142, 143, 145, 151, 152, 153, 164, 175, 190, 213
– control by fines removal, for pilot scale crystallizers   180–182
cumulative function variable   2
cycling phenomenon, as undesired effect of fines destruction   182–183

d

deconvolution   13–14
– direct inversion using nonnegativity constraint   14
– iterative methods   15
– Philips Twomey inversion method   14–15
delta-mode model predictive control   198–199
density function variable   2
desupersaturation curves   65, 86, 88
diaphragm valve   223
diastereomers   23
digital process refractometers (DPRs)   71, 72, 78
dilution system and CSD   17, 18
direct inversion, using nonnegativity constraint   14
discrete-time models versus continuous-time models   142
distributed control system (DCS)   119, 122, 196
distributed-parameter models versus lumped-parameter models   141–142
draft-tube-baffle (DTB) crystallizer   178–179, 197, 209, 213
– double effect crystallizer with external elutriation leg   212
– reactive   214
– single effect evaporative crystallizer with thermal vapor recompression   211
dynamic models versus steady-state models   142
dynamic optimization   161, 162, 171, 172
– online   163–168

e

eccentric disk rotary control valve   222
edge detection   43–44
empirical modeling   191

f

fines density measuring instrument (FDMI)   180
fines removal   175
– CSD control by, for pilot scale crystallizers   180–182
– cycling phenomenon as undesired effect of   182–183
– and dissolution   109, 118–119, 179
– by heat dissolution   175–176
– in industrial practice   178–180
– mixed suspension mixed product removal (MSMPR) crystallizer with   177–178
fines trap (FT)   176, 180
first-principles modeling   190, 191
– versus empirical modeling   141
fmincon function   154
focused beam reflectance measurement (FBRM)   21, 35, 37, 83, 107, 119, 180
– advantages and limitations   26–27
– application examples
– – different impurity level effect   23–24
– – improved downstream processing   24–25
– – metastable zone width (MSZW) and solubility   21–22
– – nucleation kinetics   24
– – polymorph transformations   22–23
– – process control   25–26
– – seed effectiveness   22
– measurement principle   21
forced circulation crystallizer   204
– single effect
– – with mechanical vapor recompression   208
– – steam heated   206
– – with thermal vapor recompression   207
– triple effect
– – evaporative crystallizer with washing thickener   210

forward light scattering 7–8
– deconvolution 13–14
– – direct inversion using nonnegativity constraint 14
– – iterative methods 15
– – Philips Twomey inversion method 14–15
– laser diffraction 8–17
– – application for monitoring and control of industrial crystallization processes 17–19
– – principles 8–10
– multiple scattering 16
– scatter theory 10–12
– – anomalous diffraction 12
– – Fraunhofer diffraction 13
– – generalized Lorenz–Mie theory 12
– shape effects 15–16
Fourier transform (FT) spectrometers 81
Fraunhofer diffraction 13
frequency histogram 5
function block diagram (FBD) 122

g
geometrical optics 12

h
hydroquinone 46

i
imaging 35–36, 108
– application for crystallization process monitoring 46–48
– literature overview 36–39
– sensor design 39–40
– – camera system and resolution 42–43
– – image analysis 43–46
– – optics and illumination 40–42
– – statistics 46
industrial crystallizers design and control 203–204
– control devices 222–224
– draft-tube-baffle crystallizer 209, *211, 212*, 213
– – Oslo growth-type crystallizer 213, 215, *216, 217*, 218
– forced circulation crystallizer 204–209
– process variables in crystallizer operation
– – batch operations 219
– – continuous operation 218–219
– sensors 219
– – crystal size 221–222
– – density 220–221
– – level 220

infrared absorption 81, *82*
in-line measurement 59
in-line process refractometer 71
– accuracy 74
– application example in crystallization 76
– – seeding point and supersaturation control in sugar vacuum pan 77–79
– concentration determination 74
– features and benefits 76
– measurement principle 72–73
– process sensor 75–76
– process temperature compensation factor 75
input constraints 193
iterative methods 15

l
Lambert–Beer law 82, 84
laser diffraction 8–17, 35, 36–37
– application for monitoring and control of industrial crystallization processes 17–19
– principles 8–10
light absorbance 29
light emitting diode (LED) 72, 76
linear models versus nonlinear models 142
Lorenz–Mie theory, generalized 12

m
Maxwell equations 12
mean gray intensity (MGI) 37
metastable zone width (MSZW) and solubility 21–22, 60–65, 132–137
Michelson interferometer 81
Mie scattering theory 7
mixed suspension mixed product removal (MSMPR) crystallizer 204
– with fines removal 177–178
model-based object recognition 45–46
model predictive control (MPC) 185
– advantages and disadvantages of 187–189
– approach for designing and implementing 189
– for batch crystallization 168–172
– constraints 193
– of crystallization processes 197–198
– delta-mode 198–199
– implementation 196
– optimization 193
– performance index 192
– process modeling 190–192
– receding horizon principle 185–187
– state estimation 194–196

m

model predictive control (MPC) (contd.)
– system structure 196
– turning 193–194
moving horizon principle. See receding horizon principle
multiple-input multiple-output (MIMO) systems 187
multiple scattering 16

n

nephelometers 52, 53, 56
nonlinear control 161, 164, 168, 172
nucleation 63, 127, 128, 130, 131, 135, 136
– kinetics 24
– measurement 51–52

o

object linking and embedding for process control (OPC) 196
observer 188, 194–196
oiling-out 22
open-loop optimal control 142, 143, 149, 150, 151, 155–157, 168
Oslo growth-type crystallizer 213, 215, 216, 217, 218
– double-stage cooling crystallization system 216
– single-stage reactive crystallizer 217
output constraints 193

p

partial least-squares regression (PLSR) 83, 98
particle size distribution (PSD) 1–4, 21, 26, 30
– characterization, based on mass distribution 5–6
– moments 4–5
particle vision and measurement (PVM) 36, 41
performance index 185, 186, 192
phase transitions detection, with ultrasound 66–67
Philips Twomey inversion method 14–15
photodetectors 52,
pinch valve 223–224
plug valve 222
polymorphism 22–23, 87–88, 97
population balance equation (PBE) 151, 152, 161
prediction horizon 194
principal component regression (PCR) 83, 98
process analytical technology (PAT) 107
process image analyzer (PIA) 36, 41
programmable logic controller (PLC) system 119, 120, 196
proportional-integral-derivative (PID) control 120, 122, 187
pseudo-random binary signals (PRBSs) 191

r

Raman scattering 94
Raman shift 95, 96, 97, 98, 100, 101
Raman signal 95, 98
Raman spectroscopy 93
– applications
– – amorphous content quantification 100–101
– – liquid phase composition monitoring 99–100
– – solid-phase composition monitoring 99
– calibration 95
– – multivariate approaches 98–99
– – univariate approaches 95–97
– factors influencing 94–95
– time-resolved 101
random-magnitude random-interval (RMRI) signal 191
rate-of-change constraints 193
Rayleigh scattering 93
receding horizon principle 185–187
refractive index 71, 73

s

saturation point 63
scaling 111
scatter theory 10–12
– anomalous diffraction 12
– Fraunhofer diffraction 13
– generalized Lorenz–Mie theory 12
seed crystals 127–128, 131, 132, 133–135, 137
seeding, as process actuator 111–112
– addition methods 114
– initial supersaturation 112
– seed mass 112–113
– seed quality and preparation procedure 113–114
– seed size and size distribution 113
seeding technique, in batch crystallization 127
– control 131–137
– main process parameters 128–130
– principles and phenomena 127–128
sequential function chart (SFC) 120, 121
sequential quadratic programming (SQP) optimization algorithm 154
sieving 3, 5

signal-to-noise ratio   81, 82, 89
single-input single-output (SISO) system   194, 197
slurry   31–34
Snell's law   71, 72
solubility   51–52, 53, 54, 56
solvent-mediated phase transitions   66
sound speed   59
– in-process ultrasound measurement   59–60
– measuring crystal growth rates   65–66
– phase transitions detection with ultrasound   66–67
– solubility and metastable zone width determination   60–65
state estimation   188, 190, 194–196
state-space (SS) model   190, 192
Stokes signal   93, 94
supersaturation   24, 26, 77–79, 108, 109, 110, 120, 127–129, *130*, 131, 139–140, 146, 149–50, 154, 155
– generation rate   114–116, 144–145
– initial   112

t
turbidimeter   51, 52, *53*
turbidimetry
– for crystal average size estimation   29
– – determination   29–31
– – for high solid slurry concentration   31–34
– and nephelometry   51
– – developed instruments   52–53
– – examined system   53–54
– – measurement of nucleation and solubility points   51–52
– – obtained results   54–57

u
ultrasonic attenuation   35
ultrasound, phase transitions detection with   66–67

v
V-notch ball control valve   222–223